THE ICE-AGE HISTORY OF NATIONAL PARKS IN THE ROCKY MOUNTAINS

THE ICE-AGE HISTORY OF NATIONAL PARKS IN THE
ROCKY MOUNTAINS

Scott A. Elias

Smithsonian Institution Press

Washington and London

© 1996 by the Smithsonian Institution

All rights reserved

Copy Editor and Typesetter: Princeton Editorial Associates

Designer: Alan Carter

Library of Congress Cataloging-in-Publication Data

Elias, Scott A.

The Ice-Age history of national parks in the Rocky Mountains / Scott A. Elias.

p. cm.

Includes bibliographical references (p. –) and index.

ISBN 1-56098-524-0 (paper : acid-free)

1. Paleoecology—Rocky Mountains Region. 2. Paleoecology—Pleistocene. 3. Geology,

Stratigraphic—Pleistocene. 4. Geology—Rocky Mountains Region. 5. National parks—Rocky Mountains

Region—History. I. Title.

QE720.E45 1995

551.7′92′0978—dc20   95-10213

British Library Cataloguing-in-Publication Data is available

Manufactured in the United States of America

02 01 00 99 98 97 96   5 4 3 2 1

♾ The paper used in this publication meets the minimum requirements of the American National Standard for

Information Sciences—Permanence of Paper for Printed Library Materials Z39.48-1984.

For permission to reproduce illustrations appearing in this book, please correspond directly with the owners of the

works, as listed in the individual captions. The Smithsonian Institution Press does not retain reproduction rights for

these illustrations individually, or maintain a file of addresses for photo sources.

*For my children, Clara and Daniel, whose sense of wonder
is an inspiration to their father*

CONTENTS

Chapter 8. Rocky Mountain National Park   136
Life in the Rarified Air

Chapter 9. Conclusion   157

Glossary   161

Index   165

ACKNOWLEDGMENTS

I thank the National Park Service Publications Office in Denver for financial support that allowed me to start writing this book. I thank the U.S. Geological Survey Photo Library in Denver for access to and permission to use several photographs of geologic features in the parks. I also thank U.S. Geological Survey geologists Paul Carrara, Rich Madole, and Ken Pierce for providing advice, reviews of those portions of the text dealing with geology, photographs, and encouragement.

Archaeologist Ken Cannon of the National Park Service Midwest Archaeological Center provided many useful suggestions and materials on the archaeology of Yellowstone and Grand Teton National parks. Geologist Wayne Hamilton of Yellowstone National Park offered a thoughtful review of the manuscript.

The faculty of the Institute of Arctic and Alpine Research, University of Colorado—including Nelson Caine, John Hollin, Mark Meier, Gifford Miller, Susan Short, Tom Stafford, and Mort and Joanne Turner—made helpful suggestions on first drafts of chapters and figures and also provided photographs.

Paleontologist Elaine Anderson of the Denver Museum of Natural History provided Pleistocene faunal lists for the park regions. Finally, I thank my family for their support throughout this project and for doing without me through many evenings and weekends.

In addition to the support of the National Park Service, other financial support for the preparation of this book was provided by a grant from the National Science Foundation to the University of Colorado for Long-Term Ecological Research (LTER), DEB-9211776. Support for my paleoecological research in the Rocky Mountains has been provided by the NSF-LTER grant and by grants from the National Park Service.

PART ONE

Paleoecology

Why We Need to Study Past Ecosystems

THIS BOOK DEALS WITH the prehistoric environments of the Rocky Mountains. The term *prehistory* covers several billion years on this planet. It is a time frame we all find difficult to grasp, even if we are well acquainted with some of the more spectacular episodes in prehistory, such as the age of dinosaurs. But having gotten that far, can we honestly say that we have a good understanding of what a million years means, or what percentage of total Earth history it represents?

At the nearer end of the prehistoric time scale are the events of the last few tens of thousands of years. It was during this time period that the glaciers of the last ice age advanced to cover much of the Rocky Mountains, then began to melt; humans moved into North America from Asia and, more recently, from Europe and Africa. We have come to learn something of that time, when large parts of this continent were clothed in primeval (old-growth) forests or unbroken expanses of tall-grass prairie. In an attempt to preserve some of that untamed wilderness, our government has set aside tracts of land as national parks, beginning with Yellowstone National Park, in 1872. In some cases, these parks represent the last, best examples of entire **ecosystems** in close to their primeval state.

My aim in writing this book is to provide an overview of this more recent period of prehistory for the Rocky Mountain region and of the methods used to reconstruct past environments. The book draws from the work of many scientists. I have brought together information from studies of fossil plants and animals as well as geological data, which can be used to reconstruct ancient climates. I also provide an overview of the early peoples of the Rockies, especially the first nomadic hunters that migrated south from Alaska and Canada.

Many words used in science may be unfamiliar to nonscientists. I have tried to keep these to a minimum in my narrative, but in some cases their use in unavoidable. When I use such words in this book, they are printed in boldface, and all boldfaced words are defined in the glossary at the back of the book. For instance, I could not avoid using the term **paleoecology** to refer to the science of reconstructing interactions between prehistoric plants and animals and their physical environments.

Chronologically, this book covers the late **Quaternary Period:** the last 125,000 years (Fig. I.1). During this interval, ice sheets advanced southward, covering Canada and much of the northern tier of states in the United States. Glaciers also crept down from mountaintops to fill valleys in the Rockies and Sierras. The late Quaternary interval is important because it bridges the gap between the ice-age world of prehistoric animals (some of which are now extinct) and modern environments and **biota**. It was a time of great change, in both physical environments and biological communities. It was also the time when human beings "came of age" in the world; they spread across the continents and started to become a major factor in the world's ecosystems.

The study of late Quaternary paleoecology is necessary to help us fathom modern ecosystems, because modern ecosystems are the direct result of these past events. To try to understand present-day environments without a knowledge of their history since the last ice age would be like trying to understand the plot of a long novel by reading only the last page.

You will meet few unfamiliar plants and animals in this book. With occasional notable exceptions, the flora and fauna discussed here are still growing or cavorting around on the North American landscape (albeit some more slowly than others). The ones that have become extinct since the ice age are a fascinating story unto themselves, one that we shall explore. Fossil studies show that the modern ecosystems did not spring full-blown onto the hills and valleys of our continent within the last few centuries. Rather, they are the product of that massive reshuffling of species that was brought about by the last ice age and indeed continues to this day.

The study of past ecosystems is really just a form of detective work. A police detective investigating a crime has to reconstruct the following aspects of the case:

1. What happened?
2. Who did it?

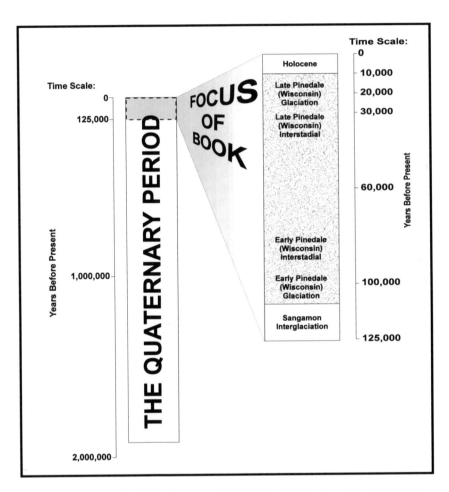

Figure I.1. The time scale of the Quaternary Period, showing the time interval that is the focus of this book: the late Quaternary, spanning the last 125,000 years.

3. How was it done?
4. When was it done?
5. Where was it done?
6. Why did it happen?

Apart from the last category, paleoecologists are basically saddled with the same questions. You might say that in fossil studies the trail of clues has grown exceedingly cold. Certainly, the suspects and witnesses are all long dead. Nevertheless, the task, although difficult, is not impossible. Moreover, it's often exciting and challenging. And, as we shall see, it is becoming more important all the time.

In order to reconstruct *what happened* in the Quaternary, we need to look at both physical and biological data. The data from the physical environment include such information as types of sediments deposited and rates of deposition, types of landforms associated with glacial and near-glacial (called periglacial) environments, and changes in lake levels. The fossils themselves tell us *who did it* and often show us *where it happened,* although we are often looking for external factors that influence the environment, such as changes in the amount of incoming solar radiation (called *insolation*). Other aspects of the physical environment make occasional appearances on the "who done it" list. These include volcanoes, earthquakes, and even the odd meteor or asteroid. The dating methods outlined in Chapter 3 provide information on *when it was done.*

Perhaps the trickiest question of all is *"How was it done?"* In other words, how have the various elements of the physical and biological world interacted on the global stage during the Quaternary? This question is the most difficult to answer, but it is also perhaps the most important, because we desperately need to know how the biological world responds to changes in the physical environment. We are inflicting our own changes on the environment of planet Earth at an ever-increasing rate. Overhunting, overfishing, pollution, and destruction of natural habitats have already wrought havoc on most ecosystems and have caused the extinction of untold numbers of species within the last few centuries. At present we find ourselves in the position of trying to understand how to sustain the remaining flora and fauna just as human activities place the natural world in ever greater jeopardy. Virtually our only means of gaining a greater understanding of how regional biotas respond to environmental change is to examine how they have responded to past changes. The current crisis thus takes paleoecology out of the arcane realm of satisfying the intellectual curiosity of a few eccentric college professors and places it squarely in the middle of worldwide efforts to save our remaining biota.

In this volume, I am focusing in on the prehistory of four national parks in the Rocky Mountains (Fig. I.2) for two reasons. One is that, as nature preserves, the parks offer excellent opportunities for scientific research in many fields, including paleontology and geology. Since the parks are relatively pristine, it is easier to compare past and present ecosystems there than in regions that have been greatly modified by human activities. I have worked on paleontological projects in Glacier, Yellowstone, and Rocky Mountain national parks and have gained the necessary familiarity with their ancient history. Another reason for my focus on the parks is that many people are interested in them and in the ecosystems they represent. This book is an attempt to fill the void between available books about modern ecosystems and those concerning bedrock geology.

Obviously, there is a large gap between the events of millions of years ago and the events of the last hundred years. The first step in bridging that gap is providing

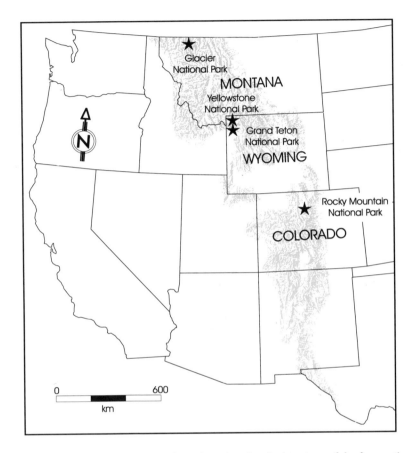

Figure I.2. Map of the Rocky Mountain region, showing the locations of the four national parks discussed in this book.

information on the hows and whys of paleoecology. Then we will examine the ancient environments, vegetation, animals, and peoples of the Rockies.

Let me issue one caution before we go any further. If you are reading this book, it is probably safe to assume that you already have some interest in fossils or archaeological artifacts. If you visit a national park and spot either fossils or artifacts, please do not touch them. The best thing to do is to mark the spot, or leave someone at the spot, and go find a park ranger. Tempting as it might be to pick up an arrowhead or a fossil bone and bring it back to a ranger station or visitor center, this would be the wrong thing to do. By removing a fossil or artifact from its original location, you would be destroying the evidence of exactly where it came from. This evidence is often vital in pinning down the age or the environmental or cultural context of the find. Fossils and artifacts that have been removed from the places where they were originally deposited cease to be useful to science; they become mere souvenirs of ancient times, stripped of the context that gives them much of their meaning.

Speaking of souvenirs, it is, of course, illegal to remove fossils and archaeological materials from national parks (along with rocks, soils, plants, and animals, except fish caught with a park-issued license). Furthermore, all research done by qualified scientists in national parks is carried out only with the permission of the National Park Service. Science officers in the parks issue permits for specific projects at specific locations.

1

QUATERNARY FOSSILS
What Are They and Where Are They Found?

Simply put, Quaternary fossils are the remains of organisms that lived during the Quaternary Period. The Quaternary Period is the most recent of the geologic periods; it is the time interval covering the last 1.7 million years and is characterized by numerous glaciations. At the turn of the century, geologists thought that there had been just four major glaciations in North America during the Quaternary Period. Now it appears that North America experienced at least 14 Quaternary glaciations. These glacial intervals are lumped together into the **Pleistocene epoch,** which began with the onset of the first glaciation (1.7 million years ago) and ended with the last glaciation. The last 10,000 years (the interval since the end of the last ice age) are called the **Holocene epoch.** This book covers events occurring at the very end of the Pleistocene and during most of the Holocene.

All of written human history has taken place in the second half of the Holocene. The last Pleistocene glaciation in North America, the **Wisconsin Glaciation,** was probably the most important force affecting the development of our modern ecosystems; it had an effect on every part of the continent, even if it did not cover the whole continent with ice. Subsequent retreat of the glacial ice opened the way for recolonization of landscapes and brought a massive reshuffling of species in every region of North America.

Vertebrate Fossils

A number of types of Quaternary fossils are commonly studied by paleontologists. Vertebrate fossils often dominate museum displays and popular literature. This notoriety is based on their high visibility and our own affinity with mammals. The skeleton of a woolly mammoth has the capacity to capture our imagination, whereas most of us have a hard time getting excited about pollen grains, **diatoms,** or bits and pieces of insects. When people think of ice-age fossils, they usually remember the fossil bones of large, extinct mammals, such as mammoths (Fig. 1.1), mastodons, and saber-toothed cats, that they have seen displayed in museums. Although these fossils are fascinating and informative, they are actually quite rare in comparison to the fossils of smaller animals and plants.

It is sometimes difficult to make paleoclimatic interpretations based on the fossil remains of large animals. One reason is that many of them probably migrated across landscapes on a regular basis and thus were able to avoid undesirable climatic conditions. On the other hand, small mammals, such as rodents, shrews,

Figure 1.1. Lower jaw of a mammoth, found in Lewis and Clark County, Montana. (Photograph by M. R. Mudge, U.S. Geological Survey.)

and rabbits (not to mention fish, reptiles, and amphibians), offer tremendous opportunities for paleoenvironmental reconstructions, because they generally remained in their small home range year-round, and their fossil remains thus provide a more reliable indication of the environmental conditions at or near the place where they are found. For instance, fossils of the collared lemming, an inhabitant of arctic tundra, have been found in late Pleistocene deposits in the American Midwest. These data provide convincing evidence that the climate of central Iowa in the late Pleistocene was quite similar to the climate found today on the Alaskan North Slope.

In addition, large mammals are rarer than small mammals in any given ecosystem (for instance, there were many more mice than mammoths wandering the North American landscape during the Pleistocene). Consequently, a given fossil assemblage may include thousands of bones (or pieces of bones) from voles, mice, and squirrels, but only a few bones from large mammals.

Although the vertebrate fossil record is important, scientists have gathered far more information from other kinds of fossils, most of which are much less glamorous; in fact, most are practically invisible to the naked eye. Among these are the pollen, stems, leaves, and fruits of plants; the **exoskeletons** of insects; the shells of snails and other mollusks; and the glassy skeletons of microscopic algae called diatoms. These types of fossils are small, but they are much more abundant in sediments than the bones of ancient mammals. In fact, just a thimbleful of lake sediment may contain thousands of pollen grains or diatoms, all wonderfully preserved down to the last detail of **microsculpture,** as viewed through a high-powered microscope. By compiling the data gathered from all of the various types of fossils from a given time period in a study region, teams of scientists are able to piece together a picture of its plant and animal life. They can then use the paleontological data to reconstruct the history of climate change. We will now focus on the major types of **microfossils** used in Quaternary research.

Fossils from Plants: Pollen and Macrofossils

Palynology is the study of pollen. It is probably the most widely used tool in terrestrial Quaternary paleoecology. Many kinds of plants produce a superabundance of pollen each year. This is especially true of wind-pollinated plants, such as conifers (evergreens). A single lodgepole pine may produce as many as 21 billion pollen grains per year. Other plants, such as insect-pollinated species, may produce only a few thousand grains per year. Pollen has an extremely durable outer wall that resists decay. The pollen grains of many plant species are light enough to float on the wind and may travel hundreds or thousands of

Figure 1.2. Researchers from the Institute of Arctic and Alpine Research at the University of Colorado obtaining a lake sediment core through the ice on Lake Dorothy, Indian Peaks Wilderness, Colorado. The ice provides a stable platform, so this method is more convenient than coring from a raft on the lake in summer. (Photograph by Peter W. Birkeland, University of Colorado.)

miles before landing. If they land in a lake, pond, or bog, they may be preserved for thousands of years.

Pollen samples are usually extracted from sediment cores taken from lakes and ponds. In the Rocky Mountain region, coring is often done in winter, through the ice on lakes and ponds (Fig. 1.2). Pollen is also preserved in buried soils, in peat, and even in glacial ice (the Greenland Ice Cap has been found to contain the pollen of spruce trees that grew thousands of miles away, in Canada).

Differences among the pollen of different species in a genus are often difficult or impossible to detect under a microscope. Because of this, pollen grains are most often identified only to the family or genus level rather than to the species (see Fig. 1.3 for an illustration of the taxonomic hierarchy). For instance, a pollen grain

Figure 1.3. The hierarchy of classification of the animal kingdom, using the example of the wolf and coyote, two species in the genus *Canis,* which is one of several genera in the family Canidae (dogs, wolves, coyotes, and foxes). The canids represent one family in the order Carnivora (meat-eating mammals). The carnivores, in turn, are one order in the class Mammalia (mammals). The mammals are one class in the phylum Chordata (animals with spinal chords). There are many phyla in the kingdom Animalia (the animal kingdom).

may be identified as originating from a pine tree, but it is not always possible to tell from which species of pine (such as white pine, jack pine, or lodgepole pine) the pollen comes. This lack of precision limits the accuracy of the interpretation of pollen data, since the ecological variability of a whole genus of plants is necessarily greater than that of individual species within that genus.

Once the pollen grains have been identified, the data are generally presented as diagrams showing pollen percentages of the total number of grains in sediment samples, plotted according to depth in a **stratigraphic column** (Fig. 1.4). Pollen diagrams from the same region and the same time interval are generally quite similar to each other but different from diagrams from other regions or times. Pollen diagrams are often divided into zones of similar pollen composition. The boundaries between zones mark transitions in regional vegetation. The proportion of pollen released into the environment depends on the number and type of plants

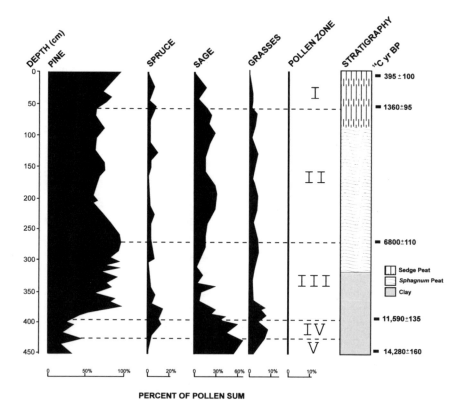

PERCENT OF POLLEN SUM

Figure 1.4. Generalized percentage pollen diagram for sites in the Rocky Mountains during the last 12,000 years. Note that the pollen zone boundaries reflect the major changes in pollen percentages through time.

present and therefore reflects the composition of regional vegetation. For instance, following the retreat of the Wisconsin Ice Sheets in North America, some regions were first colonized by low, herbaceous vegetation (grasses, sedges, and their relatives), similar to what is now found in the arctic tundra. As regional climates warmed and soils began to mature, the herbaceous vegetation gave way to coniferous forest, which in turn was invaded by hardwood trees, eventually producing the mixed deciduous–coniferous forests of today. The timing of these transitions in plant communities is usually clearly shown in regional pollen diagrams.

Some types of plants produce an overabundance of pollen, especially plants that are wind pollinated. These types of plants are often overrepresented in the regional pollen "rain" (the fallout of pollen grains from the atmosphere). Other plants, especially those that are insect pollinated, produce little pollen and are usually underrepresented in the pollen rain. Thus if a given region has a mixed forest of spruce, aspen, and willow, the pollen deposited in regional lakes and bogs will be dominated by spruce grains, with very few grains of willow or aspen. It is therefore quite difficult to determine from the pollen assemblages if aspen or willow were growing nearby. Larger plant fossils can help clarify the issue, however, since aspens and willows drop leaves, fruits, and stems as easily as spruces drop needles and cones.

Recently, palynologists have been developing statistical methods for reconstructing past environmental conditions, based on comparisons of the pollen "signature" of modern stands of vegetation with fossil pollen assemblages. By analyzing the climatic conditions under which the modern vegetation is growing and the proportions and amounts of pollen types in modern sediments from the various types of plant communities, they have been able to come up with fairly precise estimates of past climates. In addition, palynologists have used pollen diagrams to track the long-term movements of plant species and genera through the late Quaternary. Long-lived plants, such as most trees, do not spread rapidly across landscapes. Their migration in response to climate change may take centuries, but they have undergone some remarkable shifts in distribution through the past few thousand years. This may seem like a long time to us, but it represents only a small fraction of the time that a given plant species has been in existence.

The macroscopic (visible to the naked eye) remains of plants, commonly preserved in Quaternary sediments, are called **macrofossils.** Woody plants produce a great deal of potential macrofossils throughout their life cycle. As might be expected, wood is very resistant to decay in poorly oxygenated, waterlogged sediments. Specialists in fossil wood identification are able to determine species of trees by analyzing cross sections of stems. The annual growth rings in trees can provide both a history of local environments (for example, drought versus abundant precipitation, or episodes of forest fires or insect infestation) and a year-by-year chronology of those events. Macrofossils from coniferous trees also include needles and cones, both of which can frequently be identified to the species level.

Nonwoody plants, such as grasses and herbs, also produce a wide variety of macrofossils, including stems, leaves, and fruits. Peat is essentially composed of layer upon layer of undecomposed plant leaves and stems. The two principal types of peats are moss peats (frequently dominated by sphagnum mosses) and sedge peats. Moss fragments can often be specifically identified. Sphagnum mosses in bogs dominated many ice-free regions during the Pleistocene, and the combination of living and dead (peat) mosses probably represents the largest single terrestrial repository of carbon.

Plant macrofossils generally accumulate in **water-lain sediments** and reflect only local conditions, since they come from plants growing in the **catchment basin.** In order to develop a regional reconstruction of past vegetation based on macrofossils, it is necessary to study multiple sites and piece together the data from each site into a synthesis of information.

Insect Fossils

Insect fossil studies began in earnest during the 1960s and have become one of the most important sources of terrestrial data on past environments. These studies have, for the most part, employed fossil insects as **proxy data,** that is, as indirect evidence for past environmental conditions.

Beetles are the largest order of insects. They have been the main insect group studied from Quaternary sediments, and in fact they are the most diverse group of organisms on Earth, with more than one million species known to science (that's more than all of the flowering plants combined). In addition, their exoskeletons, reinforced with **chitin,** are extremely robust and are commonly preserved in large numbers in lake sediments, peats, and some other types of deposits. In most cases, beetles have quite specialized habitats that apparently have not changed appreciably during the Quaternary. This characteristic makes them excellent environmental indicators. The exoskeletons of beetles and some other insects are covered with exquisite microsculpture, enabling paleontologists to identify fossil exoskeletons to the species level in at least half of all preserved specimens, even though insect exoskeletons are most often broken up into the individual plates in fossil specimens.

Beetles are very quick to colonize a region when suitable habitats become available. They often respond more quickly than plants, which, until recently, were relied upon almost exclusively as indicators of environmental change on land. Like plant macrofossils, insect fossils are generally deposited in the catchment basin in which the specimens lived. Thus they provide a record of local conditions, in contrast to pollen, which can be carried many miles on winds and often gives a more regional "signal."

Studies of insect fossils in two-million-year-old deposits from the high arctic have failed to show any significant evidence of either species evolution or extinction. Beetle species have apparently remained constant for as many as several million generations.

Insect fossils are generally extracted from organic-rich lake or pond sediments or peats. Ancient stream flotsam, deposited in **fluvial sediments** and later exposed along stream banks, is often a rich source of insect fossils.

Insect fossil data are usually presented as minimum numbers of individuals for each species identified. Paleoclimatic reconstructions are generally made on the basis of the climatic conditions in the region where the species in a given assemblage can be found living together today, that is, the climate of the region where their modern distributions overlap.

Diatoms, Ostracodes, and Others

Diatoms, ostracodes, and other organisms can provide information on ancient lake conditions. This, in turn, can often be tied directly to climate, because the size and water quality of lakes are controlled largely by the balance between evaporation, precipitation, and local ground-water conditions. The fossil organisms in lakes reflect changes in regional climate, as those changes affect lake water temperature, salinity, alkalinity, and other factors. However, sometimes local factors, such as the presence of springs, **alkaline sediments,** or proximity to glacial ice, may overwhelm the regional climatic signal.

The group of primitive plants lumped together under the common name *algae* is quite diverse and extremely successful, having colonized most regions of the globe, including the land, fresh water, and the seas. The most diverse group of algae are the diatoms. They also happen to be the best-preserved algal group in Quaternary sediments because they produce cell walls made of silica, which resists decomposition. Because diatoms preserve well and grow in large numbers in both lakes and the oceans, their siliceous remains can reach concentrations of one billion per cubic centimeter of sediment!

Diatoms are not often used for paleoclimatic reconstructions because their presence in a body of water is more dependent on water quality than on regional climate. Nevertheless, they can be quite useful in revealing a number of ancient conditions. For instance, some freshwater diatoms live only in shallow water, whereas others live only in deep water. Some of those living in shallow water cling

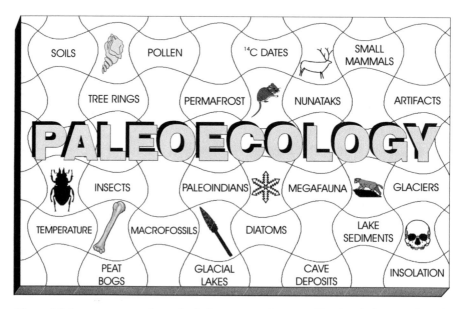

Figure 1.5. In some ways, paleoecological reconstruction resembles the piecing together of a jigsaw puzzle.

only to certain types of substrates (rocks, sand grains, mud, or plants). Other important factors controlling the type of diatoms found in a lake include the salinity, nutrient level, and acidity or alkalinity of the water.

Ostracods are tiny, bivalved crustaceans that live in fresh or salt water. Their shells, or carapaces, are reinforced with calcite (crystalline calcium carbonate), which dissolves in nonalkaline waters. Their fossil shells are therefore found in nonacidic sediments, such as **marl.** The abundance and diversity of ostracods in a given lake are usually dependent on salinity, amount of oxygen, acidity–alkalinity, water depth, and food availability. Some are found only in lakes; others prefer ponds; still others live in running water.

A host of other "critters" in lakes and ponds leave behind fossils that can serve in paleoenvironmental reconstructions. Among these are freshwater sponges, water fleas, and snails (both terrestrial and aquatic). Freshwater sponges filter water through their pores, straining out bacteria for food. This makes them delicate monitors of water quality. The "skeleton" of sponges is made up of silica spicules, which preserve well in lake sediments and can often be identified to the species level.

Water fleas, or cladocerans, are one of the most important groups of tiny crustaceans that live in fresh water. Like diatoms, water fleas are divided into groups that live in shallow and deep waters. They too are seldom used in paleoclimatic reconstructions, but their fossils can tell a great deal about past lake conditions, including nutrient levels, salinity, and acidity–alkalinity. The exoskeletons of water fleas are made of chitin.

A variety of these aquatic plants and animals can be found in abundance in water-lain sediments. It may seem redundant to study more than one or two types of fossils, since many of them provide overlapping information on water quality and water depth. However, paleoecologists don't look at it that way. They consider each piece of fossil data to be relevant, because each fossil group usually has some unique data to contribute, and the combined information always makes for a better, more sharply defined picture of past environments. If nothing else, different data sets serve as corroborative evidence, confirming the information provided by some of the more commonly studied groups.

As we have seen, paleoecology is basically detective work, with fossils serving as witnesses to past environments. The fossils have a lot to say if you can understand their language. The process of paleoenvironmental reconstruction is a lot like fitting together pieces of a jigsaw puzzle. The more pieces you have, the more successful you'll be (Fig. 1.5).

Suggested Reading

Birks, H. J. B., and Birks, H. H. 1980. *Quaternary Palaeoecology.* London: Edward Arnold. 289 pp.

Eicher, D. L. 1976. *Geologic Time.* Englewood Cliffs, New Jersey: Prentice Hall, 150 pp.

Elias, S. A. 1994. *Quaternary Insects and Their Environments.* Washington, D.C.: Smithsonian Institution Press. 284 pp.

Warner, B. G. (ed.). 1990. *Methods in Quaternary Ecology.* Geoscience Canada Reprint Series No. 5. St. John's, Newfoundland: Geological Association of Canada. 170 pp.

2

THE REPOSITORIES OF ECOLOGICAL HISTORY

Where Are Ice-Age Fossils Found?

When I mention to people that I work on insect fossils, they often assume that I have to hack away at outcrops of shale, looking for faint impressions of fossils in the stone. In fact, Quaternary fossils are most often found in mud, not in stone. That is because most Quaternary sediments are **unconsolidated,** that is, they are loose soils, sands, clays, and other matter that have not yet turned to stone (given a few million years more under pressure and/or great heat, they may). Quaternary fossils can also be found in peat moss (the kind used for gardening). Peat is composed of waterlogged plants that accumulate in bogs because of extremely slow decomposition. Given a few million years under the right conditions, peat turns to coal.

Pre-Quaternary fossils in bedrock generally fall into one of two categories: they are either mineral replacements of the original organic matter or impressions of ancient plants and animals. In both cases, the original organic matter has long since decomposed. However, most Quaternary fossils are the actual remains of plants and animals that were preserved before decomposition set in. As mentioned previously, this type of preservation occurs in peat bogs. It also takes place at the bottom of lakes and ponds, where the sediments are low in or without oxygen (oxygen supports aerobic bacteria that decompose and destroy the structure of organisms). In very dry environments, such as desert caves and rockshelters, the remains of plants and animals gradually dry out and may become "mummified" fossils that can last tens of thousands of years.

Quaternary fossils are found in a wide variety of sediments. We will consider only terrestrial and freshwater sediments in this chapter. These materials are quite familiar to everyone. They are the soil in your garden, the smelly mud at the bottom of a pond, and the dust blowing into your house during a windstorm.

Types of Sediments Containing Fossils

Mineral sediments, such as gravel, sand, silt, and clay, accumulate in four main types of deposits: eolian, alluvial, colluvial, and lacustrine. Pleistocene glaciers and ice sheets created vast quantities of sediments as they pulverized the rock over which they moved. Glaciers act as giant conveyor belts, transporting rocks and finer debris both on top of and inside the layers of moving ice. Glacial deposits range from huge boulders to fine silts and clays that get left behind when the ice retreats.

Eolian Deposits

Eolian deposits are wind-blown silts or sands. Loess is a deposit of relatively uniform, fine sediment (mostly silt), that was transported to the deposition site by wind. Most loess deposits are relatively inorganic and contain few fossils. Loess mantles large regions of North America, including Alaska, Washington, and Nebraska, in depths up to several tens of meters.

Alluvial Deposits

Alluvial or stream deposits are created by moving water; they include gravels, sands, and silts. Stream deposits often include pockets of organic debris, which accumulated in backwaters, pools, or low-energy side channels. These deposits can be a treasure trove of fossils because they consist of stream flotsam, which may accumulate in large quantities over a short time. When river channels shift, they sometimes abandon a side channel, which then becomes an oxbow lake. Once an oxbow lake forms, it rapidly fills in with vegetation and eventually becomes dry land. In the process, however, the water-lain organic detritus in the oxbow becomes part of the fossil record.

Colluvial Deposits

Colluvial deposits are created by the downslope movement of sediments, often by slope wash or mudflows during storms. In the alpine zone of the Rockies, saturated

soils creep downslope over frozen ground or are frost-heaved downslope in processes called *solifluction*. Solifluction is a cold-region phenomenon that was more
widespread during the Pleistocene than it is now. Slope wash, mudflows, and
solifluction transport organic materials that lie on or near the surface to the
bottom of a slope. Slope wash acts in the minutes or hours during and after a
storm. The action of mudflows may stretch to days or weeks. Deposition by
solifluction takes decades to centuries. However, the ultimate results are similar,
namely, the movement of organic materials from hillsides down to the streams,
ponds, and lakes where they are redeposited.

   Organic deposits are also contained in buried soils. These occasionally include
usable fossils, notably pollen. However, most organic matter in soils is more or less
decomposed because of exposure to oxygen, so fossil preservation is generally not
as good as in water-lain sediments. The other problem with buried soils is that the
process of soil development does little to concentrate fossils, so they are few and far
between.

Lacustrine Deposits

Water tends to concentrate organic detritus (such as stream flotsam and organic
debris in lakes) into recognizable layers that eventually produce more specimens
per cubic centimeter than neighboring upland soils (Fig. 2.1). Therefore, the
principal types of sediments containing abundant fossils are those that are laid

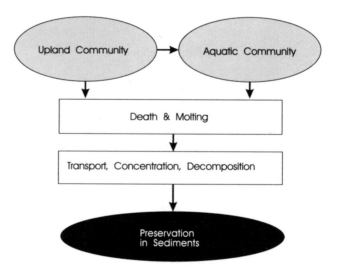

Figure 2.1. Summary of the preservation of organisms in the fossil record of a catchment
basin.

down in water. Water in a basin, such as a pond or lake, may collect large amounts of plant and animal matter from its watershed. Dead insects, leaves, seeds, and twigs are carried to ponds and lakes by streams. Insects, pollen, and small seeds are blown in by the wind or washed in by streams. Some animals (both large and small) simply fall into the water and drown or are washed downstream after they die (Fig. 2.2).

Not all lake sediments are alike. Sediments rich in calcium carbonate, known as marls, preserve fossil bones and the shells of mollusks and ostracods better than sediments that are more **acidic.** In some lakes, the layers of sediments laid down each year are poorly defined, whereas in others there are clearly marked couplets of sediment (winter and summer sediment layers) for each year. These annual layers are called *varves.* Varved sediments allow researchers to count back year by

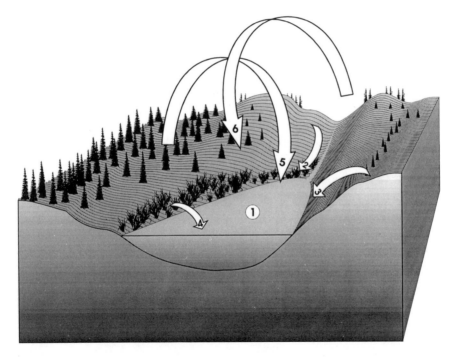

Figure 2.2. Generalized drawing of a mountain watershed, showing sources of potential fossils that may be deposited in a lake. 1, plants and animals that live in the lake; 2, plants and animals carried into the lake by a stream (including stream-dwellers and organisms that fall into the stream); 3, plants and animals transported to the lake by slope wash, erosion, or solifluction; 4, lakeshore plants and animals that fall into the water; 5, microfossils (mostly pollen and spores) from nearby plant communities, transported to the lake by wind; 6, microfossils (mostly pollen and spores) from distant plant communities, transported to the lake by wind.

year in the fossil record of a lake, leading to incredible accuracy in the dating of events.

In addition to fossil preservation, a great deal of information about past environments can be gleaned from the study of the physical and chemical properties of lake sediments. The size of the particles in the sediment (gravel, sand, silt, and clay) reveals whether they were deposited in a low-energy environment (in the lake, or from a slow-moving stream) or a high-energy environment (from a river, a flooding creek, or a beach gravel redeposited after a severe storm). Most deposition takes place in low-energy environments, whereas erosion takes place in high-energy environments.

Peat Bogs

The two principal types of peats are moss peats (frequently dominated by *Sphagnum* mosses) and sedge peats. Moss peats accumulate in bogs, whereas sedge peats accumulate in fens. Peat bogs are fed by precipitation and are acidic; they are also poor in nutrients. The water in fens often comes from a combination of precipitation and ground water; they are often less acidic than bogs and richer in nutrients.

Peat forms when the remains of aquatic or semiaquatic plants in a bog or fen accumulate faster than they decompose. This process is helped along by the water in the bog or fen. Plant remains that sink to the bottom of the water decompose much more slowly than plant remains in the open air. Decomposition takes place readily in warm, well-oxygenated environments, but the bottom of a bog is often cold and oxygen poor. Bogs may form as a part of **ecological succession** from a lake or pond to a meadow. Many mountain meadows are filled-in ponds or lakes. The modern vegetation of these meadows is growing over layers of peat, sometimes several meters thick.

Bogs cover large regions of Canada, Alaska, and Siberia today. During the Pleistocene, bogs developed in many unglaciated regions, leaving substantial peat deposits behind.

Cave Deposits

In a generic sense, the term *cave* covers subterranean caverns of all sizes; the term *rockshelter* is more appropriate for most shallow caves inhabited by prehistoric animals and people in the Rocky Mountains and elsewhere. Rockshelters are shallow enough to allow considerable sunlight and air circulation, even at their deepest point, whereas caves may extend many miles underground and include regions of complete darkness and little exchange of air with the outside world. I use the term *cave* here in the more generic sense.

For our purposes, cave deposits are most important as a source of vertebrate fossils. Caves are essentially closed systems, limited to walls, floor, and roof. They have a beginning (cave formation) and an end (collapse of the cave roof or infilling by outside sediments). Such temporal boundaries mean that the sediments (and associated fossils) that collect in a cave represent a discrete time interval; although some caves persist for many thousands of years, many may remain open for only a few centuries or millennia. This is in contrast to many other types of sediment deposition, such as that in large lakes or rivers, where sediments may accumulate for tens or hundreds of thousands of years.

Accumulation of sediments in a cave begins when an opening forms to the outside world. Most caves that have produced abundant vertebrate bones are small, shallow caves that are close enough to the surface to have substantial contact with world outside the subterranean cavern.

There are five sources for the bones found in cave deposits (Fig. 2.3):

1. Animals carried in by predators (including both ancient and modern people).
2. Fecal pellets of animals or birds frequenting or living in the cave.

Figure 2.3. Generalized section of a cave, showing potential sources of fossil bones: 1, predators that carry prey animals into the cave; 2, owl and other raptor pellets; 3, predators attracted by carrion that fall into the cave through large vents; 4, animals that fall into the cave through vents; 5, animals that live in the cave. (Modified from Andrews, 1990.)

3. Predators or scavengers that are attracted to carrion, then become trapped after entering the cave.
4. Animals that fall into the cave through cracks or other openings from the surface.
5. Cave-dwelling animals.

Other than specialized cave dwellers, most animals will only live in caves that afford easy access to the outside world. Pleistocene caves provided shelter for numerous animals, including humans. When they died in caves, their remains were often preserved in cave-floor sediments. However, cave researchers have concluded that most bones in cave deposits are from animals that accidentally fell in, rather than from actual cave dwellers. A good example of this kind of cave is Natural Trap Cave, east of Yellowstone, in northern Wyoming. It has a small opening at the top, which widens to form oversteepened walls inside. Such caves often contain deposits with an abundance of carnivores (mountain lions, wolves, and other meat-eating mammals). It is thought that the carnivores are attracted to the smell of carrion emanating from the opening at the top of the cave and then enter (or fall into) the cave themselves and are unable to get out.

Predators, including birds of prey, often carry prey animals into the open mouths of caves. The bones of the prey are then deposited in the cave as refuse left over from the predators' meals, in regurgitated pellets, or in the predators' scat.

Taphonomy: How Fossils Become Preserved

The process by which a living organism becomes preserved in the fossil record is called **taphonomy.** Paleontologists have devoted entire careers to the study of taphonomy, or "the laws of burial". Figures 2.1 and 2.2 present brief summaries of the taphonomic processes for a lake or pond in a catchment basin. Although the figures are an oversimplification of a very complex set of events, they show the basic sources of potential fossil material on a living landscape and the general processes that lead to the preservation of plants and animals in sediments.

Most organisms that end up in the bottoms of lakes and ponds spent their lives either in the water or in close proximity to it. In fact, it makes intuitive sense that more aquatic and riparian (shore-dwelling) organisms are preserved in water-lain sediments than members of upland species, simply because the odds are smaller that the remains of an upland creature will make their way into the lake. The remains of many upland creatures, both plants and animals, decompose on the surface and are not preserved as fossils. Even the boniest, most hard-bodied creatures will decompose if left to rot on a hillside. Some large upland animals die near or in the water or are washed downslope by rains, so their carcasses end up in water-lain sediments, but on the whole we have a better understanding of what life was like in or near the water and a poorer knowledge of ancient life on dry hillsides.

Another confounding factor is that some aquatic invertebrates (such as caddis-fly larvae and ostracods) go through several stages of development before reaching maturity. With each new stage, they shed their old exoskeleton or shell. This molting process creates several sets of potential fossils for each individual.

Microfossils, including many types of pollen grains, float through the air and travel quite readily for many miles. This characteristic ensures that the pollen that rains out of the sky is representative of a broad region, including uplands and lowlands. Wind-pollinated plants, such as conifers, are overrepresented in the pollen records of lakes, because even though aquatic plants are assured of pollen deposition in the lake, the wind-pollinated plants produce orders of magnitude more pollen, so they simply overwhelm the aquatic plants in the pollen rain.

Upland insects land in lakes and streams by accidentally falling in or by being blown into the water during flight. However, the proportion of upland and aquatic or shore-dwelling insects in most fossil assemblages is surprisingly well balanced.

Vertebrate skeletons (except for those of fish) are only occasionally preserved in water-lain sediments. Most good vertebrate fossil deposits have been found in caves or other natural traps. The taphonomy of vertebrate fossils is summarized in Figure 2.4. Bones may be broken or cracked when the animal dies (either by the impact of a fall into a natural trap or by predators); shortly after death, they may be affected by gnawing or breakage from scavengers or by trampling by other animals. Once a carcass is reduced to an accumulation of bones, chemical and physical weathering may come into play, and such factors as frost heaving, soil erosion, mudflows, and the like may move the bones around and mix them with other bones in a deposit.

Under some circumstances, then, fossils may lie relatively undisturbed in sediments through many thousands of years and beyond. To the untrained observer, these deposits may look like worthless piles of mud or layers of peat best sent to a gardener. To the paleontologist, however, these repositories represent a bank vault, ready and waiting to offer up a treasure trove of fascinating clues to the history of life on this planet.

Destruction of Fossil Resources

There are a wide variety of sources for Quaternary fossils. These fossils are all around us, literally underfoot. Unfortunately, many valuable sources of Quaternary fossils are being exploited by people for other purposes. For example, peat is

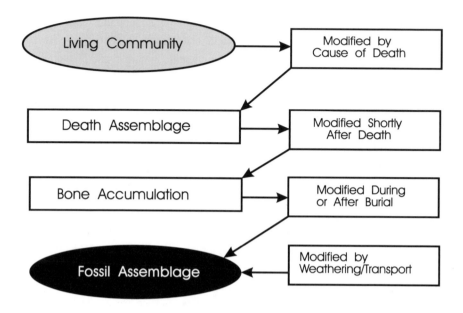

Figure 2.4. Summary of bone taphonomy processes. (Modified from Andrews, 1990.)

mined from now-shrinking deposits in North America, Europe, and the former Soviet Union. Peat is used both as organic matter to improve garden soils and, in some places, as a fuel for heating and cooking (albeit not a very good one). Ancient fluvial deposits are mined as sources of sand and gravel or precious metals, such as gold. Although such mining activities have brought many Pleistocene fossils to light through the years, they have devoured many more. Ancient mammoth ivory has been used extensively for jewelry, buttons, and even piano keys and billiard balls. Artifacts from archaeological sites, such as arrow- and spearheads, pottery, and stone carvings, are carried away by "pot hunters." Yet these materials are the keys that unlock the book of prehistory; it is a shame when they are reduced to just another traded commodity or collectible curiosity.

Suggested Reading

Andrews, P. 1990. *Owls, Caves, and Fossils.* Chicago: University of Chicago Press. 231 pp.

Birks, H. J. B., and Birks, H. H. 1980. *Quaternary Palaeoecology.* London: Edward Arnold. 289 pp.

3

DATING PAST EVENTS

Once we have assembled bits and pieces of information from various types of fossils, the next step in reconstructing past environments is to fit the fossil data into a time frame or chronology. This procedure may be approached in a number of different ways, depending on what types of materials are available for dating and the interval of time we wish to study (Fig. 3.1). Accurate dating is essential to paleoecology; without it, it is impossible to determine the rates at which past environmental change took place (for example, did a climatic warming begin rapidly or more slowly?). Accurate dating also makes it possible to determine whether past events took place at the same time across a broad region or whether those events were unrelated to each other.

In order to explain most Quaternary dating methods, we must make a brief excursion into the highly technical fields of high-energy physics and organic chemistry. This is the realm of **isotopes, ions,** and isomers. In this chapter I have attempted to provide an overview of the major methods, explaining enough to allow the reader to grasp the basic idea without delving into the details with which dating specialists must deal on a day-to-day basis. It always helps to understand the principles behind the technology one relies on, even if that understanding is somewhat rudimentary. For additional information, please consult the suggested readings at the end of the chapter.

29

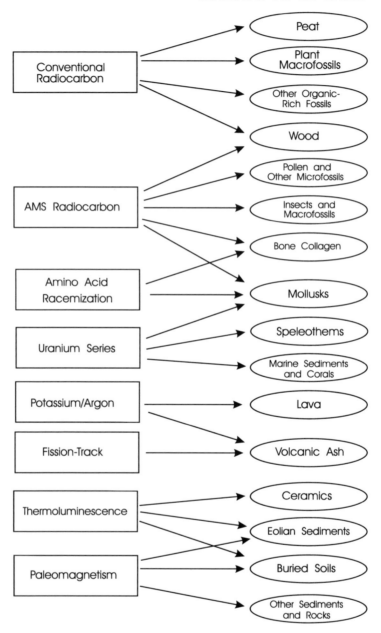

Figure 3.1. Summary of radiometric, chemical, and paleomagnetic dating methods, showing the types of fossils and other materials dated by the methods.

Types of Dating Methods

Dating methods fall into four categories: (1) radiometric methods; (2) paleo-magnetic methods; (3) chemical methods; and (4) biological methods. Radiometric methods measure the radioactive decay of unstable isotopes of elements, including isotopes of carbon, potassium, and uranium. Paleomagnetic methods measure changes in the magnetic field of the earth as revealed in changes in the polarity of magnetically charged minerals. Chemical methods measure time-dependent changes in certain chemicals, including those in fossils. Finally, biological methods measure the growth of long-lived plants, especially trees (as expressed in tree rings) and certain species of lichens.

Each of the various dating methods is useful over a certain time span (Fig. 3.2). Some are used to date events only within the last few thousand years. Others only begin to date events 10,000–15,000 years old or older.

The principal radiometric methods used in Quaternary studies are based on the decay of radiocarbon to stable carbon (^{14}C to ^{12}C), potassium/argon dating, and uranium series dating. I discuss each method in turn.

Radiocarbon Dating

Radiocarbon dating is the most widely used dating method for late Quaternary fossils. The ^{14}C isotope of carbon is continuously being created in the upper

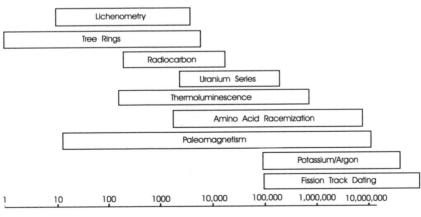

Figure 3.2. Summary of radiometric, chemical, paleomagnetic, and biological dating methods, showing the time ranges each method is capable of measuring.

atmosphere through the bombardment of nitrogen atoms by neutrons from cosmic rays from the sun (Fig. 3.3). The neutrons react with stable nitrogen atoms (^{14}N) to create a radiocarbon atom and a hydrogen atom. The radiocarbon atoms rapidly combine with oxygen to form $^{14}CO_2$, which diffuses down through the atmosphere and is taken up by plants in **photosynthesis.** The plants store the ^{14}C atoms in their tissues, along with much larger quantities of stable (^{12}C and ^{13}C) carbon. The ^{14}C atoms are continuously taken in by plants, and secondarily by the animals who eat the plants, throughout their lifetime. Even predators at the top of the food chain (such as eagles, lions, and wolves) have ^{14}C in their tissues in the same proportion to ^{12}C as is found in the atmosphere.

Once plants and animals die, they stop taking in ^{14}C, and the ^{14}C in their bodies begins to be depleted. Since the ^{14}C atom is radioactive, it begins to decay back to nitrogen (^{14}N). The rate of this decay was first determined by Willard Libby in

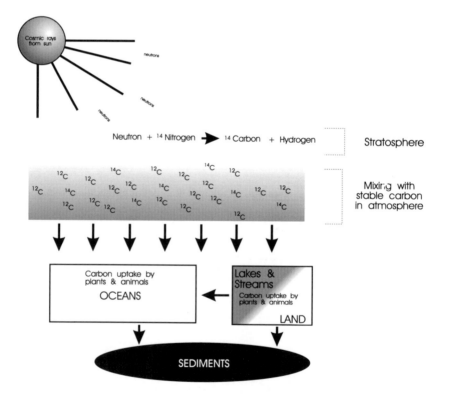

Figure 3.3. Synopsis of how radiocarbon is created and spreads in the atmosphere, its uptake by living organisms, and its final deposition in sediments. The arrows show the direction of movement of radiocarbon in the system.

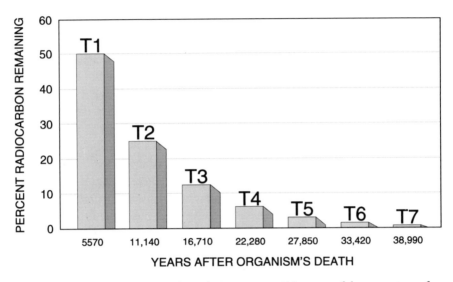

Figure 3.4. Decay of radiocarbon through time, expressed in terms of the percentage of original radiocarbon atoms remaining in a sample through seven half-lives.

1955. Half of the ^{14}C atoms will have decayed to nitrogen within 5570 years. This time interval is called the *half-life* of the radioactive decay. Following the progression (Fig. 3.4), then, by 11,140 years after the death of an organism, only 25% of the original ^{14}C content will remain. By 16,710 years after death, only 12.5% of the ^{14}C will remain. By 22,280 years after death, 6.25% of the ^{14}C will be left. This exponential decrease in the amount of remaining radiocarbon can be measured to approximately 10 half-lives, or 55,700 years before present (B.P.). In practice, ^{14}C dating is reliable only for samples younger than 40,000 yr B.P. There is so little radiocarbon left in an older fossil that it becomes virtually impossible to measure accurately. Fossils older than 40,000–45,000 years yield radiocarbon ages of 40,000–50,000 yr B.P., whether they are 50,000 years old or 50,000,000 years old. Radiocarbon laboratories designate such samples as, for instance, "greater than 40,000 yr B.P."; this is then taken as the sample's minimum age, since it could be much older.

There are now two methods of obtaining radiocarbon ages from a sample. For large samples of organic material (twigs, logs, or blocks of peat), conventional methods may be used, including either gas-counting or liquid scintillation. In the gas-counting method, the sample is burned and the gases emitted from burning (such as carbon dioxide and methane) are placed in a chamber with a detector that counts emissions of **beta** (β) particles (electrons given off in the radioactive decay process). The liquid scintillation technique involves converting the organic component of a sample into a liquid (benzene or some other organic liquid). This liquid is placed in a chamber that detects scintillations, the minute flashes of light given off when a β particle is emitted from the liquid.

Within the last decade, the development of the accelerator mass spectrometer (AMS) method has allowed much smaller samples to be radiocarbon dated. Whereas the conventional methods require samples containing at least several grams of carbon to yield reliable dates, the AMS method is able to date minuscule samples, such as individual seeds, insect parts, and tiny lumps of charcoal. For instance, the AMS method was used recently to obtain a ^{14}C age on single strands of fabric from the Shroud of Turin, which was claimed by some to have been the burial shroud of Christ. In fact, AMS dates from three separate laboratories showed that the Shroud of Turin was made during medieval times. Instead of measuring β particle emissions, which are an indirect measure of the amount of ^{14}C in a sample, the AMS method measures the actual abundance of the carbon isotopes (^{14}C, ^{13}C, and ^{12}C) in a sample. The atoms are accelerated in a **cyclotron** or tandem accelerator chamber to extremely high velocities. Then they are passed through a magnetic field, which separates the different isotopes, allowing them to be distinguished from each other and counted individually.

Potassium/Argon Dating

Potassium/argon (K/Ar) dating is based on the radioactive decay of ^{40}K to ^{40}Ar. Potassium commonly occurs in two stable isotopes (^{39}K and ^{41}K), plus small amounts (0.012%) of the unstable isotope, ^{40}K. In comparison to radiocarbon, ^{40}K decays very slowly. The half-life of the decay of ^{40}K to ^{40}Ar is 1.31 billion years. Argon is a gas. It can be driven out of a rock sample by heating. The K/Ar method is used to date volcanic rocks in which all initial argon was driven out when the lava was in a molten state. As time passes, ^{40}Ar builds up in the lava again from the decay of radioactive potassium. By measuring both the ^{40}K and ^{40}Ar content of a lava sample, the ratio of potassium to argon can be found and the number of half-lives expended can be calculated. However, because the radioactive decay is so slow, this method is only useful for samples greater than 100,000 years old.

Uranium Series Dating

Uranium series dating is based on the radioactive decay of uranium, either the ^{238}U or the ^{235}U isotope. The ultimate product of uranium decay is stable lead (^{206}Pb or ^{207}Pb). However, the process involves several intermediate isotopes of various elements, called daughter products of the radioactive decay. For instance, one of the principal daughter products of the decay of ^{238}U is the thorium isotope ^{230}Th, which has a half-life of 75,200 years. One of the daughter products of the decay of

^{235}U is the protactinium isotope ^{231}Pa, which has a half-life of 32,400 years. These two daughter products are the main isotopes studied in uranium series dating. One important feature of the thorium and protactinium isotopes is that they do not easily dissolve in water. This means that they **precipitate** out of solution and collect in sediments. Because the concentration of uranium in sea water is a constant, its accumulation rate in sea sediments is also known, and the age of the sediment can be estimated, based on the percentage of uranium that has decayed into the daughter products.

The useful dating range for ^{238}U/^{230}Th is 10,000–350,000 years. The useful dating range for ^{235}U/^{231}Pa is 5,000–150,000 years. Uranium series dating is used mostly for marine fossils, such as corals. It is also used to date terrestrial fossils or features containing carbonates, such as speleothems (stalactites and stalagmites) from caves and travertine layers deposited by springs, such as Mammoth Hot Springs in Yellowstone National Park.

Thermoluminescence

Thermoluminescence (TL) is the light emitted from a mineral crystal when it is heated following the mineral's exposure to radiation. Electrons that are produced by radioactive decay (β particles) become trapped in the crystal matrix of minerals. When the mineral is heated, the electrons escape and give off light. The longer the mineral has been in the ground, collecting free electrons, the more light it will give off when heated. Since heating discharges the electrons from the crystals, it sets the TL "clock" back to zero. This makes TL useful for dating archaeological artifacts such as ceramic pottery. The clay minerals used to make a pot will discharge their TL completely when they are fired in a kiln. Any subsequent measurable TL can then be used to date the time elapsed since the ceramics were made.

Thermoluminescence only works for sediments or artifacts that are buried in complete darkness. This is because sunlight also empties the minerals of their TL and resets their TL clock. Buried loess (windblown silt) deposits and buried soils have been shown to retain their TL "charge" until they are dug up to be sampled. The useful age range of TL dates spans the interval of roughly the last million years; the accuracy of TL dating needs further refinement.

Chemical Methods

One of the principal chemical dating methods is amino acid dating. Amino acids are the building-block molecules of proteins. They are made by all living organisms. The structure of amino acid molecules is asymmetric or lopsided: the left-hand side of the molecule does not match the shape of the right-hand side. This asymmetric property creates an optical effect when polarized light is passed through amino acid molecules. One configuration, or isomer, of the amino acid will rotate the plane of light to the left. This isomer is called the *levo* (Greek for "left") form. Nearly all amino acids in living organisms are in a *levo* configuration. The other isomer causes light to rotate to the right. This form is called the *dextro* (Greek for "right") form. After an organism dies, the amino acids in its body, which began as *levo*, or "L," isomers, slowly begin to change to *dextro*, or "D," isomers. This process is not unlike radioactive decay, except that it represents a chemical change that stops when the isomers reach an equilibrium point, whereas radioactive decay proceeds in one direction until the supply of unstable isotopes is exhausted.

The process by which amino acids change from one isomer to the other is called *racemization*. Amino acid racemization occurs at different rates in different organisms and is also greatly affected (as are most chemical reactions) by temperature changes. Still, amino acid racemization is a useful technique for dating one set of fossils relative to another set of fossils of the same species. Corrections can be made for fossils from different localities that have been in sediments of different temperatures.

The other principal chemical dating method is the chemical aspect of ash dating, or *tephrochronology*. The minerals that make up volcanic ash vary from one volcano to another and from one volcanic eruption to another. Because of this, each eruption produces volcanic ash, or *tephra*, with a unique chemical composition or "signature." This chemical signature is not used to date the timing of the eruption (there are other methods for doing that), but once a specific volcanic ash has been dated, the ash itself becomes a very useful tool for correlating regional sediments. For instance, the Norse people (also known as Vikings) landed on Iceland about A.D. 900, in the same year as an Icelandic volcano erupted (Icelandic volcanos are *very* active). The ash from that eruption is called the Landnám tephra (Landnám is the Old Norse word for *landing*). Wherever that tephra is found preserved in sediments on Iceland, it provides a convenient horizon that marks the timing of human arrival on the island.

Tephras are mostly dated in one of three ways. Late Pleistocene and Holocene ashes have been dated by radiocarbon dating of associated organic material, such as charred wood fragments or peat layers in bogs that are directly overlain by the ash. Ashes older than about 40,000 yr B.P. are dated by two other methods. If the mineral content of the ash contains sufficient potassium, it may be K/Ar dated. Ashes that contain volcanic glass or minerals such as zircons may be dated by the fission-track method. Fission-track dating is based on the decay of uranium isotopes. As these decay, they emit radiation in the form of **alpha** (α) and β **particles.** The energy released in this process causes two nuclear fragments to be thrown out into the surrounding material. The resulting gouge marks, or paths in the minerals or glass, are called *fission tracks* (tracks caused by nuclear fission). These tracks are extremely short (on the order of a hundredth of a millimeter). After the volcanic glass or mineral has been polished and chemically etched to bring out the tracks, they can be counted under a microscope. Then the glass is heated, which causes the surface to smooth over, or anneal, as it melts. Once the ancient fission tracks are gone, the glass is exposed to radiation under controlled conditions in the laboratory. The new radiation produces new fission tracks as the result of fission of ^{235}U. The number of newly created ^{235}U fission tracks is proportional to the uranium content, enabling the ^{238}U content of the glass sample to be calculated. So, when all is said and done, fission-track dating is just an indirect method of uranium series dating, allowing the researcher to date very small glass shards from volcanic eruptions.

Other radiometric and chemical dating methods, which we will not discuss, include the obsidian hydration, lead-210 (^{210}Pb), and beryllium isotope methods.

Biological Methods

The biological dating methods are more "user friendly." That is, they are easier to understand and don't involve much in the way of high-powered physics or chemistry (as a biologist, I'll admit to a healthy share of bias in this statement). There are two principal biological dating methods. One is tree-ring counting, or *dendrochronology;* the other is the measurement of the growth of certain lichens, or *lichenometry.*

Dendrochronology is based on the fact that, in temperate climates, trees lay down annual rings as they grow. Each ring is made up of a broad, light-colored band that represents growth in the spring and summer and a narrow, dark-colored band that represents lessened activity in the fall. Trees tend to grow broader rings during "good" years (years with adequate warmth, moisture, and nutrients) and narrower rings during "bad" years (years of drought, disease, and cold summers). The pattern of broad and narrow rings is repeated in trees growing throughout forest regions. This allows tree rings to be matched, or correlated, from tree to tree, and even to dead trees, a process that in turn allows tree-ring chronologies to be developed over greater lengths of time than the life of individual trees. Of course, tree-ring researchers seek out very long-lived trees, such as bristlecone pines, from which to take cores for their studies. The cores are extracted from small holes drilled into the tree by a tool called an increment corer. These holes do not usually harm the tree; they are quickly filled in with resin, which seals the wound. Bristlecones and some other conifers may live several thousand years. By piecing together tree-ring chronologies from living and dead trees (trees whose lives

Figure 3.5. *Alectoria* lichen growing on a boulder in a glacial moraine, Baffin Island. (Photograph by Gifford H. Miller.)

overlapped at some point in the past), chronologies have been extended back throughout the Holocene, or the last 10,000 years. Tree-ring dating is so reliable and accurate that it has been used as a method of calibrating radiocarbon dating. This is done by chiseling out pieces of wood from individual rings of known age and ^{14}C dating the wood.

Lichenometry is based on the assumption that certain species of lichens grow at very slow, predictable rates. When these lichens are found growing on such objects as boulders on a glacial moraine, they can be used as an indirect way of dating glacial movements, by dating how long those boulders have been in place. The idea here is that when a rock gets caught up in glacial ice, all the lichens growing on it are scoured off or die as they are buried below the surface of the ice. Once the glacier stops moving or retreats, it leaves the boulder behind, and the clean surface of the boulder rapidly becomes colonized by new lichens.

Many species of lichens grow radially; that is, they grow through a series of expanding circles rather than stretching along a line. By measuring the diameter of radial-growth lichens, it is possible to estimate their age (Fig. 3.5). An aid to this procedure is to measure the size of lichens growing on an object of known age. One favorite source of this information is lichens growing on gravestones, particularly in Europe, where gravestones date back many centuries. One of the most commonly used lichens for lichenometry is the **crustose lichen,** *Rhizocarpon geographicum.* Lichens can live for several thousand years, especially in cold climates. One problem with lichenometry is that lichen growth tends to slow down with age. Another is that it is exceedingly difficult to identify lichens accurately (they are, after all, a combination of an alga and a fungus, growing in **symbiosis**).

Many tools are used to obtain the ages of sediments and fossils. Some are more precise than others, and all are applicable to only certain segments of the Quaternary time line. Often, they are used in combination. For instance, a radiocarbon age may be obtained from organic materials in silt, and a TL age obtained from the silt itself. However, none of these dating methods is cheap. The current charge for a conventional ^{14}C date is about $250, and an AMS radiocarbon date can cost more than $1000, depending on how much work is required to prepare the sample for dating; not many Quaternary researchers can afford to obtain more than a few dates from a given site. A common practice for organic-rich samples that are thought to be less than 45,000 years old is to obtain a radiocarbon date from a basal sample, one from near the middle of the stratigraphic section, and one from the top. When possible, researchers try to get additional dates on horizons representing times of environmental change, so that the timing of those changes can be pinned down more precisely.

Suggested Reading

Berglund, B. E. (ed.). 1986. Dating methods. Chapters 14–19 in *Handbook of Holocene Palaeoecology and Palaeohydrology.* New York: John Wiley & Sons. 869 pp.

Bradley, R. S. 1985. Dating methods I and II. Chapters 3 and 4 in *Quaternary Paleoclimatology: Methods of Paleoclimatic Reconstruction.* Boston: Allen & Unwin. 472 pp.

Fritts, H. C. 1976. *Tree Rings and Climate.* New York: Academic Press. 567 pp.

Schweingruber, F. H. 1988. *Tree Rings: Basics and Applications of Dendrochronology.* Boston: Reidel. 276 pp.

4

PUTTING IT ALL TOGETHER

We have covered a lot of ground in the last three chapters. I have tried to bring the reader up to date on types of Quaternary fossils, where they are found, and how deposits are dated. In this chapter, I provide a summary of how these data are combined to reconstruct past environments. The job of synthesizing the data into a meaningful reconstruction of past environments is often the most difficult aspect of Quaternary science, but it can also be the most satisfying. After months of sieving samples, peering down a microscope, or boiling sediments in chemical baths for a pollen preparation, it is refreshing to step back from the mundane tasks and discuss the "big picture" with colleagues. The process may take weeks or months (sometimes even years) to complete, and the initial attempts to fit the data together may end in the painful decision to go out and get more or better samples before any of the important questions can be answered. The process takes patience, perseverance, and a broad outlook.

The study of ancient climates is a necessary step in unraveling ancient ecosystems, because the physical environment plays a strong role in defining ecosystems. The first level of paleoclimatic reconstruction is the process of planning the research project. This begins with a determination of which research questions are important and answerable (or at least which hypotheses can be tested). Once the research plan has been drawn up (and permission obtained from the National Park Service if the work is to be done in a national park), the scientist can begin collecting data. Data collection involves field work to collect samples, followed by laboratory analyses of samples.

The second level of research involves converting the raw data from the laboratory into paleoclimatic data. This process ranges from simple, qualitative approaches to complex, quantitative approaches. The simplest approach is to find the modern distributions of the species in a fossil assemblage and then determine the geographic region where the modern distributions of the species in question overlap. Then the modern climate of that region can be used as an estimate of what conditions were at the time and place associated with the fossil assemblage. More sophisticated approaches involve statistical transformations of the fossil data, using mathematical formulas based on modern observations to fine-tune estimates of paleoclimates.

The third level of paleoclimate reconstruction involves combining a series of local studies into a regional synthesis. This type of study attempts to describe regional climatic patterns (such as average seasonal temperatures) over a given time interval and then proceeds to compare the patterns derived from the proxy data with theoretical climate models, such as models that reconstruct general circulation patterns in the atmosphere. Data are best synthesized by a team of researchers, each of whom approaches the research questions from a different angle or discipline (such as climatology, paleontology, or ecology). This type of interdisciplinary research (work that crosses the boundaries between scientific disciplines) is the most valuable, for it produces the most coherent, tightly focused answers, tested from many different perspectives. The nature of interdisciplinary research is that the scientists cooperate to arrive at an answer to one question using different means, often derived from different disciplines.

Imagination is one aspect of paleoecological research that, surprisingly, is quite essential. You may be able to assemble all the facts and figures from a fossil assemblage, but unless you can recreate a prehistoric scene in your imagination, you probably will not put the data together in a very meaningful way. This does not mean that our analyses are just a bunch of daydreams; far from it! Rather, it means that researchers have to combine their knowledge of how things work in the living world with the assembled body of fossil data in order to develop more than a superficial understanding of past events. This process involves the principle of *uniformitarianism*, formulated by British geologist Charles Lyell in 1830. This principle is summed up by the saying "the present is the key to the past." It might equally well be said that the past and present are interlocking parts of the whole, each an inseparable key to the understanding of the other. All of our modern animals and plants are just the latest generation of species that began in the distant past. If we are to understand them well enough to preserve today's populations, we need to study their history . . . their ancestral lineages that trace back hundreds of thousands of years. Thus the past becomes the key to the future.

The remainder of this chapter provides an overview of how the various bits and pieces of ancient biological and physical data are combined, or synthesized, to form the major paleoenvironmental reconstructions that are sometimes called "the big picture."

Reconstructing Physical Environments

The history of the physical environment is an integral part of paleoecology, because plants and animals operate in the physical world and respond as much to changes in the physical environment as they do to biotic interactions (e.g., competition for resources, predation, and parasitism). The physical environment can be broken down into three parts: the land (the geosphere), the water (the hydrosphere), and the air (the atmosphere). Obviously, the three elements interact continuously with each other. Nearly all of the energy that drives these interactions ultimately comes from the sun.

Changes in the physical environment are recorded in the features of ancient landforms. In cold regions, such features as **glacial moraines** and ancient **permafrost** features are evidence of past glacial and **periglacial environments.** These features are presently active above **treeline** in the Rockies. Permanently frozen ground leaves several types of features on a landscape, including ice wedges, patterned ground, and stone stripes. Ice wedges form when the ground contracts in cold temperatures (–15 to –20°C, or 5 to –4°F), causing cracks at the surface. Patterned ground is another feature of permafrost landscapes; there are two types. One is caused by the melting of bodies of ice in the ground. This type occurs at high latitudes and at the margins of active glaciers. The other type of patterned ground is created by frost action in the soil. This type of patterned ground occurs in mountainous regions, even far south of the continuous permafrost zone (Fig. 4.1). Finally, frost-heaving over long periods of time tends to sort the material on or near the surface, pushing cobbles and pebbles into ridges that form a polygonal net or series of rings. On steep hillsides, gravity forces the shape of the sorted material into a series of stripes, hence the name stone stripes.

Information on past glaciations can be derived through the study of shifts in the elevation of snowlines and glaciation thresholds. The boundary on a glacier between regions where winter snows melt off in summer and regions of permanent snow is called the *equilibrium line altitude* or ELA. Above this line, snow continues to accumulate. It compresses under its own weight and forms ice, adding to the

Figure 4.1. Patterned ground, Lee Ridge area, Glacier National Park. (Photograph by Paul Carrara, U.S. Geological Survey.)

mass of the glacier. Below that line, snow does not accumulate year after year, and ice does not build up.

ELAs tend to fall on a climatic boundary that corresponds to regions in which summer temperatures average at the freezing point of water, 0°C (32°F). However, different climatic conditions seem to control ELAs in different regions. Obviously, precipitation plays a major role. No matter how cold it is, if there is no new snow, new ice will not form. During glacial intervals, temperatures were depressed to the extent that snow accumulated not just on mountaintops, but across much of the high-latitude regions, eventually forming ice sheets that covered most of the northern regions and spread south to the midlatitudes. In mountains, glaciers advanced when the amount of ice buildup above the ELA was greater than the amount of ice melt-out below the ELA. In the Rocky Mountains, ELAs during the late Wisconsin Glaciation were as much as 1000 m (3280 ft) lower than they are today.

Past movements of glaciers are difficult to trace. Glaciers transport debris to their margins as they move, and where the glaciers from two mountain valleys come together, they frequently leave a pile of debris along their junction as well. When the ice retreats, moraines are left behind as testimony to past ice advances. Moraines consist of mounds and ridges of glacial debris, unsorted silt, sand, pebbles, cobbles, and boulders. They provide obvious evidence of glaciation. There are two major problems, however, with trying to reconstruct past glacial events from morainal evidence. One is that each new glacial advance tends to obliterate the evidence from previous events. So the glacier that advances the farthest down a valley will grind up and redeposit the moraines of all previous shorter advances. The second major problem is in trying to obtain an accurate date for past glacial events. Two methods are commonly used. One is to obtain a ^{14}C age from soils that develop on moraines. This age will only be a minimum age, because it may take several centuries to develop a good organic soil on a moraine. Moraines have also been dated by obtaining ^{14}C ages on logs incorporated into their sediments. Another method is to date the time when the rocks in the moraine stopped moving by using lichenometry. As we have already seen, this technique is not very accurate, although it works better on more recent moraines (those less than a few thousand years old) than on older ones.

Other methods of reconstructing past physical environments include studies of past lake levels and the study of the physical properties of lake sediments. Past lake levels can be deduced from ancient beach ridges or terraces that indicate the altitude of past shorelines. The quantity and type of sediments found in lake sediments can reveal information on changes in the lake itself and on changes in local environments, such as the timing of episodes of soil erosion, input of sediments from nearby glaciers, and other features.

Reconstructing Climate Change

Fluctuations in the amount of insolation (*incoming solar radiation*) are the most important cause of large-scale changes in Earth's climate during the Quaternary. In other words, variations in the intensity and timing of heat from the sun are the most likely cause of the glacial–interglacial cycles. This solar variable was neatly described by the Serbian cosmologist and mathematician Milutin Milankovitch in 1938. Three major components of the Earth's orbit about the sun contribute to changes in our climate (Fig. 4.2). First, the Earth's spin on its axis is wobbly, much like a spinning top that starts to wobble after it slows down. This wobble amounts to a variation of up to 23.5° to either side of the axis (Fig. 4.2A). The amount of tilt in the Earth's rotation affects the amount of sunlight striking the different parts of the globe. The greater the tilt, the stronger the difference in seasons (i.e., more tilt equals sharper differences between summer and winter temperatures). The full range of the tilt (from left of center to right of center and back again) is expressed over a period of 41,000 years.

As a result of a wobble in the Earth's spin, the position of the Earth on its elliptical path changes relative to the time of year. For instance, in Figure 4.2B,

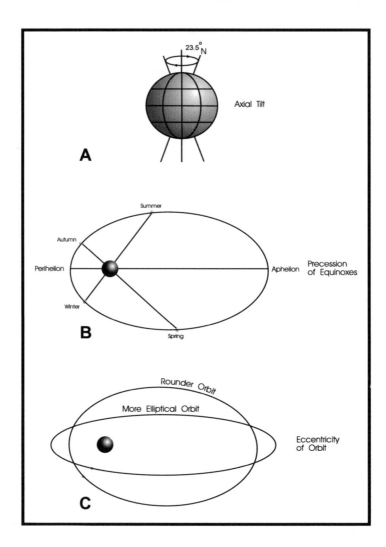

Figure 4.2. The three major components of the Earth's orbit about the sun that contribute to changes in our climate, as described in the Milankovitch theory of ice ages.

autumn and winter occur in the northern hemisphere when the Earth is relatively close to the sun, whereas summer and spring occur when the Earth is relatively far from the sun. This phenomenon is called the *precession of equinoxes*. The cycle of equinox precession takes 23,000 years to complete. In the growth of continental ice sheets, summer temperatures are probably more important than winter. Throughout the Quaternary period, high-latitude winters have been cold enough to allow snow to accumulate. It is when the *summers* are cold (i.e., summers that occur when the sun is at its farthest point in the Earth's orbit) that the snows of previous winters do not melt completely. When this process continues for centuries, ice sheets begin to form.

Finally, the shape of the Earth's orbit also changes. At one extreme, the orbit is more circular, so that each season receives about the same amount of insolation. At the other extreme, the orbital ellipse is stretched longer, exaggerating the differences between seasons (Fig. 4.2C). The eccentricity of the Earth's orbit also proceeds through a long cycle, which takes 100,000 years.

Major glacial events in the Quaternary have occurred when the phases of axial tilt, precession of equinoxes, and eccentricity of orbit are all lined up to give the northern hemisphere the least amount of summer insolation. Conversely, major **interglacial** periods have occurred when the three factors line up to give the northern hemisphere the greatest amount of summer insolation. The last major convergence of factors giving us maximum summer warmth occurred between 11,000 and 9000 years ago, at the transition between the last glaciation and the current interglacial, the Holocene.

There are other factors involved in climate change, such as volcanic eruptions and large-scale dust storms. For the most part, these phenomena trigger short-term climate change (on the scale of years to decades); however, they may interact with Milankovitch factors to draw global climate over the threshold into another ice age.

Paleoclimatic reconstructions involve piecing together data from fossil and other physical evidence and combining it with a chronology based on stratigraphy and various types of dating. As with all of science, there are inherent problems in paleoclimatic reconstruction. One is that each type of proxy data responds to climate in its own way and at its own rate. For instance, some types of plants (especially trees) may take centuries or millennia to respond fully to a major climatic change, whereas insects may respond to the same change within a few years or decades. In this scenario, tree pollen and insect fossils from the same lake sediment samples tell very different stories. Such was the case for samples from the end of the last glaciation in Britain. The fossil insect faunas showed that summer temperatures were very close to modern levels, but the pollen records were still registering arctic tundra plants. Eventually, the conifers and then the hardwood trees arrived, but not until many centuries later.

Another problem with paleoclimatic reconstructions based on proxy data is that some data are more or less continuously deposited, whereas other data are discontinuous or spotty. For instance, a lake basin may collect sediments every year for thousands of years, whereas an adjacent glacier leaves moraines that date only to one century.

Reconstructing Ecosystems

Reconstructing ancient climates may seem difficult, but it is relatively straightforward compared to reconstructing past ecosystems. The reason is simple: plants and animals behave in much more complicated and unpredictable ways than sediments, glaciers, and frozen ground. If individual species responded independently to their environment, paleoecology would be simpler. Unfortunately, the species are constantly interacting with each other in ways that are very hard to see in a fossil record. An example of this problem is the extinction of large mammals at the end of the Pleistocene. We may never know whether the animals were the victims of environmental change or human hunting or a combination of the two.

Like paleoclimate studies, paleoecological studies can be divided into first-level or descriptive approaches and second-level or statistical approaches. The second-level approach is gaining in popularity and success. It employs numerical and computer techniques and can sometimes reveal subtle patterns in the data that would otherwise be missed. Statistical methods can also be used to standardize the paleoecological reconstructions of several researchers working in a given region.

In order to reconstruct past ecosystems from fossil data, paleoecologists must develop studies that sample well-preserved fossils in the appropriate biological groups from the right time interval to answer the research question. Most paleoecologists are trained to identify fossils in one or a few groups of plants or animals. There are very few generalists in paleoecology, because virtually no one has the time and energy to learn to identify all the types of fossils found in a deposit. For instance, in order to identify fossil insects from sites in North America, it is necessary to develop a good working knowledge of about 100 different families of beetles, ants, and other groups!

Once a group of fossils has been identified from an assemblage, it is necessary to explore the ecological requirements of their modern counterparts (assuming the species in the fossil assemblage have not become extinct). Each species is adapted to a range of environmental tolerances and forms a part of a biological community. The factors that control its position within that community are described as its *niche*. The gathering of data on individual species' ecological requirements provides a good deal of paleoecological information, but a clearer picture can be obtained by reconstructing whole communities, because the community itself occupies a certain niche within a geographic region, and that niche is more narrowly defined than the niches of individual species (Fig. 4.3). For example, an alpine tundra community may be able to exist only on isolated mountaintops in the Rockies, even though several species that are a part of that community are able to live in a variety of habitats away from such slopes.

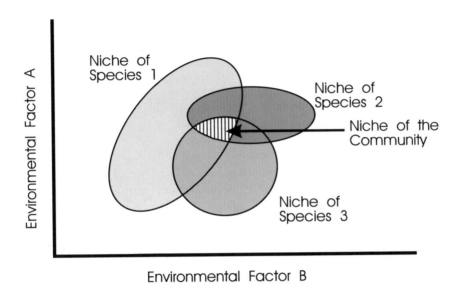

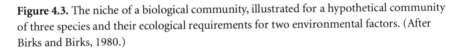

Figure 4.3. The niche of a biological community, illustrated for a hypothetical community of three species and their ecological requirements for two environmental factors. (After Birks and Birks, 1980.)

This principle is good in theory, but fossil records hardly ever preserve whole communities intact. However, if enough elements in a community are found in a fossil assemblage, comparisons with modern communities may be possible. This process is also strengthened by the presence of *indicator species,* that is, species that are strongly indicative of certain communities. For instance, mastodons appear to have lived only in close proximity to coniferous forests (conifer needles were a major part of their diet). Wherever mastodon fossils are found, it is reasonable to assume that coniferous forests and an associated boreal climate were also present.

Finally, the reconstruction of past ecosystems is based on the compilation of fossil assemblages from a number of regional sites representing various communities. Ancient ecosystems are described with the aid of analogous modern ecosystems. Comparing and contrasting modern analogues with ancient biotas fills in some of the gaps while at the same time showing differences in the structure and function of past and present ecosystems. No ancient ecosystem was exactly like any modern ecosystem. This is because ecosystems are constrained by the conditions of the physical environment, which has changed continuously through time. In addition, past communities were made up of unique mixtures of species. Even though there can be similarities between communities through time, species migrate, become established in new regions, and die out in others. All of this is in response to changing environments, competition between species, and changing resource availability.

Archaeology: How People Fit into the Picture

For the most part, humans had less impact on North American ecosystems than they did on ecosystems in Europe and Asia, at least until the Europeans arrived on the scene, beginning in 1492. This fact is helpful to paleoecologists working on North American sites. In contrast, our European counterparts are often faced with the dilemma of trying to separate human modifications of past landscapes from natural changes brought about by climate change or other factors. In other words, anthropogenic (human-induced) effects on European landscapes date back several thousand years, confounding attempts to reconstruct natural environments and ecosystems.

One of the biggest remaining mysteries in the study of North American ecosystems concerns the mass extinction of large mammals, or megafauna, at the end of the last ice age, 11,000 years ago. One school of thought holds that most of the North American megafauna were driven to extinction by early hunters (Paleoindians) shortly after they arrived from Asia via the **Bering Land Bridge.** Another theory holds that the megafauna became extinct because their habitats were greatly disrupted as environments changed at the end of the Ice Age. Some scientists believe that the megafauna were already on their way out because of environmental factors and that the Paleoindians merely delivered the coup de grâce. I will discuss this issue more fully in Chapter 6.

The fields of archaeology and paleoecology collaborate under the banner of *geoarchaeology*. Research in this field attempts to fit archaeology more closely into a paleoenvironmental scenario. Fossil data are collected from the archaeological site or from adjacent natural deposits or both. Moreover, geoarchaeologists use fossil data (rather than just archaeological artifacts) to help develop an understanding of prehistoric peoples and their ways.

Suggested Reading

Birks, H. J. B., and Birks, H. H. 1980. *Quaternary Palaeoecology.* London: Edward Arnold. 289 pp.

Bradley, R. S. 1985. Non-marine geological evidence. Chapter 7 in *Quaternary Paleoclimatology: Methods of Paleoclimatic Reconstruction.* Boston: Allen & Unwin. 472 pp.

Dearing, J. A., and Foster, I. D. L. 1986. Lake sediments and palaeohydrological studies. In Berglund, B. E. (ed.), *Handbook of Holocene Palaeoecology and Palaeohydrology.* New York: Academic Press, pp. 67–90.

Digerfeldt, G. 1986. Studies on past lake-level fluctuations. In Berglund, B. E. (ed.), *Handbook of Holocene Palaeoecology and Palaeohydrology.* New York: Academic Press, pp. 127–131.

Dixon, E. J. 1993. *Quest for the Origins of the First Americans.* Albuquerque: University of New Mexico Press. 154 pp.

Holliday, V. T. (ed.). 1992. *Soils in Archaeology: Landscape Evolution and Human Occupation.* Washington, D.C.: Smithsonian Institution Press. 254 pp.

Rapp, G., Jr., and Gifford, J. A. (eds.). 1985. *Archaeological Geology.* New Haven: Yale University Press. 435 pp.

PART TWO

Ancient Life and Environments of the National Parks of the Rocky Mountains

THE ROCKY MOUNTAINS ARE in many ways the backbone of North America, dividing east from west. This division takes many forms. The Rockies divide the continent's watersheds: the precipitation that falls on the east side of the Rockies ends up in the Atlantic Ocean, while the precipitation that falls on the west side ends up in the Pacific Ocean. During the late Pleistocene, the Rocky Mountain regions of Canada and the regions farther west were almost engulfed in the **Cordilleran Ice Sheet,** while most of Canada east of the Rockies and the north-central and northeastern United States were covered by the **Laurentide Ice Sheet.** The divide between the two ice sheets lay east of the Rockies, with the two ice bodies meeting near the U.S.-Canadian border in eastern Montana (Fig. II.1). As we will see in subsequent chapters, the Rockies had glaciers of their own during the late Pleistocene. Although these glaciers did not spread out onto the surrounding lowlands to any great extent, they certainly had impacts on the mountains and their biota.

Another very important division brought about by the Rocky Mountains is the separation of the flora and fauna of the East from those of the West. Many species of plants and animals are found only on one side of the Continental Divide. The great Midwestern prairies end in the eastern foothills of the Rockies. The moun-

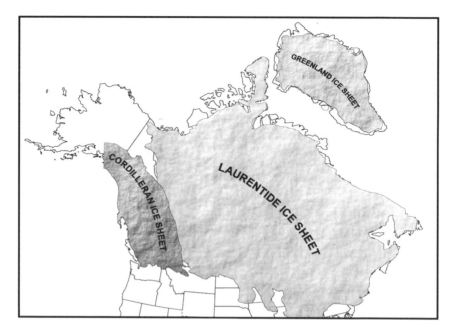

Figure II.1. Map of northwestern United States, showing the two ice sheets that came together near the U.S.-Canadian border in eastern Montana.

tain slopes serve as a narrow "sea" of coniferous forest, extending down from the boreal forest "ocean" of Canada. At the tops of the mountains, small "islands" of alpine tundra persist above the "sea" of forest below. When you travel up the slopes of the Rockies, you pass through a similar set of life zones as if you were traveling north to the arctic. You progress from the grasslands of the plains through montane forests of pine and other conifers, up to the subalpine forests of spruce and fir, finally reaching the tundra on top. Similarly, a trip north leads you from the temperate forests or prairies into the boreal forests of Canada, ending up in the arctic tundra zone in the high latitudes. If you were to travel even farther north, you would end up in the high arctic, where bare rocks and ice dominate the landscape. Similarly, the highest peaks of the Rockies are tipped with permanent snow and ice, crusted onto bare rocks on steep slopes.

During the late Pleistocene, the zones (both the latitudinal zones of the continent and the altitudinal zones of the Rockies) were pushed southward and downslope. Arctic tundra became established south of the continental ice sheets in the Midwestern United States. Boreal forest grew south of the tundra zone. In the Rockies, the boundary between alpine tundra and subalpine forest (upper treeline) pushed down the mountain slopes by hundreds of meters, and coniferous forest crept out onto the plains in some places. At the end of the last glaciation, the various groups of biological communities moved back north and back upslope to their present elevational zones. However, these postglacial shifts did not come about through the movements of whole communities. As with any major environmental change, each species responded individually, and the composition of the various communities changed, sometimes quite dramatically. Rather than an orderly transition up slopes or from south to north, the biological shifts that took place at the end of the ice age were far more chaotic. It was as if the captain of the great ship *Pleistocene* got on the public address system and announced, "Ladies and gentlemen, I regret to inform you that the ice age is now over. Furthermore, we are sinking, and the lifeboats are too small to handle all of you at once, so I'm afraid it's *every man for himself!* Good luck, and we hope to see you again during the next glaciation."

It is important to remember that, for most of the last two million years, Pleistocene glaciations have dominated the high- and mid-latitude regions of the world. The interglacial periods, such as the one in which we live, are thought to have been far shorter than the glaciations. In other words, the warmth that has allowed the development of our modern ecosystems and their interglacial predecessors has been the exception, not the rule, for most of the last two million years. For perhaps as much as 90% of the Quaternary, the world has been held in the grip of glacial ice. This is a hard concept for us to grasp, because human civilization has grown up entirely within the current interglacial period, the Holocene. We know nothing else. Our ancient ancestors, those who hunted woolly mammoth and sought shelter in caves, have left us no written record of what it was like to live in glacial times.

Given that ice ages are the norm and interglacial warmth is the exception, it follows that the cold-hardy biota of the mountains and high latitudes occupied more real estate through most of the Quaternary Period than they do today. Conversely, the warmth-loving biota are now enjoying a brief respite from the severe restrictions placed on them by those seemingly interminable ice ages. The shift from glacial to interglacial climates took place at least fourteen times in the Pleistocene. Because of the nature of the fossil record (younger glaciations tend to wipe out the evidence of previous events), our best evidence concerns the transition from the Wisconsin Glaciation to the Holocene, about 10,000 years ago. As with all large-scale environmental transformations, this last one wrought havoc with the existing biological communities. We will explore these changes, and what they mean for modern ecosystems, as the story of each park region unfolds.

5

GLACIER NATIONAL PARK
Northern Forests and Rivers of Ice

Glacier National Park (Fig. 5.1) is the only park in the Rocky Mountain region with a climate suitable for maintaining substantial glaciers since the end of the last glaciation. This climate exists there for two reasons. One is that the park is situated far enough north and its mountains are high enough to keep cool in summer. The northern boundary of the park lies at the U.S.-Canadian border. The other reason is that the mountains in the park capture significant precipitation from moist, Pacific air moving inland. The west side of Glacier National Park catches the lion's share of Pacific moisture. The difference between the wetter climate of the western slope and the drier climate of the eastern slope affects the modern-day vegetation of the park, and it made quite a difference in the size, shape, and location of late Wisconsin glaciers.

The park lies in the northwest corner of Montana (Fig. 5.2) and contains the northernmost portion of the Rocky Mountains in the United States. The glacial geology of the park has been more thoroughly studied than that in other Rocky Mountain regions of Montana. The impressive glaciers that existed in northwestern Montana at the turn of the twentieth century, along with the impressive mountain scenery and wildlife, led to the creation of the park. In some ways, the way in which the park got its name is quite fortuitous. Not many people realize that

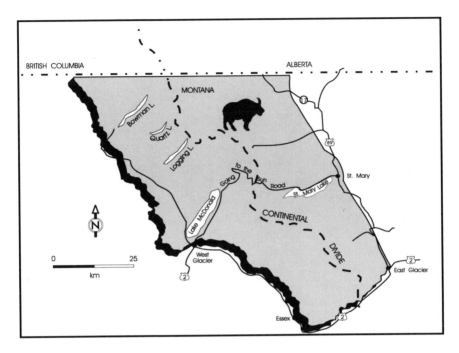

Figure 5.1. Map of Glacier National Park.

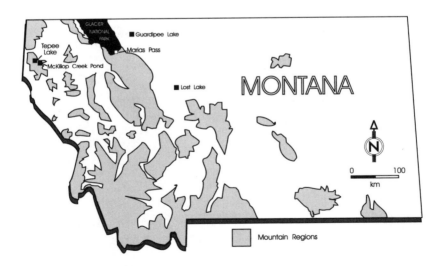

Figure 5.2. Map of Montana, showing location of Glacier National Park and several fossil localities in northwestern Montana discussed in the text. The shaded regions show the locations of major mountain ranges.

when the park was founded, the glaciers there were in the process of thinning and melting back from their extent in about 1850, when they were bigger than they had been in the last 10,000 years! Paul Carrara of the U.S. Geological Survey has determined that the mid-nineteenth-century glaciers in the park were larger than they had been since the end of the Wisconsin Glaciation. Thus the glaciers were far more impressive when George Grinnell saw them in 1885 than they are today. (Grinnell was a naturalist and magazine editor who was among the first Europeans to explore many of the glaciers in the park region, including the one bearing his name.) But the nineteenth-century glaciers were kind enough to leave us some proof of their size and shape, in the form of lateral and **terminal moraines.** Several of these moraines have been dated through the use of tree rings. The age of trees growing within the area from which the Agassiz and Jackson glaciers had retreated indicated their maximum position during the "Little Ice Age." The ice was substantial for many decades thereafter. Glaciers flowed down mountainsides, spilled over ledges, and choked lakes with icebergs. Where the ice flowed over steep terrain, it formed deep, dangerous crevasses (Fig. 5.3). The bedrock beneath the glaciers was scoured, polished, and grooved as ice and rock debris ground their way downslope (Fig. 5.4).

At about 4000 km^2 (1545 square miles), Glacier National Park is not only large but also full of rugged terrain and scenic splendor. As might be expected, nearly all of the physical features in the park have been chiseled, gouged, engraved, or

Figure 5.3. Crevasses in a glacier at Iceberg Lake, 1911. Glacial ice was far more extensive in the park at that time than it is today. (Photograph by W. C. Alden, U.S. Geological Survey.)

Figure 5.4. Glacial grooves and striae on rocks east of Vulture Peak, 1913. The grooves and striae run parallel to the direction of ice flow. (Photograph by M. R. Campbell, U.S. Geological Survey.)

trimmed by glacial ice. For instance, Lake McDonald owes its shape and size to the repeated movements of Pleistocene glaciers between the massive shoulders of Howe Ridge on the northwest and Snyder Ridge on the southeast. Sonar studies of the sediments in the lake show that the bedrock beneath it has been carved into a steep, V-shaped notch, about 270 m (885 ft) below the level of the lake (Fig. 5.5). This notch is filled with about 150 m (490 ft) of glacial sediments that have accumulated since the late Pleistocene. The series of lakes in the northwestern sector of the park are all oriented in this northeast-southwest direction (Fig. 5.1). Logging, Quartz, Bowman, and Kintla lakes are the products of glaciers that once flowed ever so slowly down their respective drainages. The mountain valleys that wind their way between the high peaks are U-shaped, the product of glacial scouring (Fig. 5.6).

The gently rolling plains east of the mountains in northern Montana were also shaped by glacial activity. They are mantled by a layer of glacial sediment, called *drift*, that was laid down by successive glaciations. It is difficult to date these drift deposits because they are all very similar in appearance and composition.

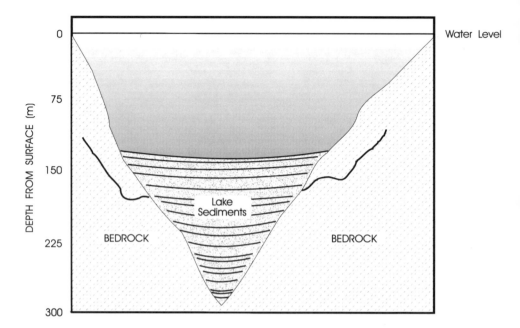

Figure 5.5. Diagram showing the accumulation of late Pinedale age sediments in the V-shaped valley overlain by Lake McDonald, based on sonar studies of the lake basin.

Let's take a closer look at the remarkable events in the Pleistocene that have shaped the park and the rest of northwestern Montana.

Glacial History: The Shaping of the Landscape

The Wisconsin Glaciation is called the Pinedale Glaciation in the Rocky Mountain region (after terminal moraines near the town of Pinedale, Wyoming). The Pinedale Glaciation began after the last (Sangamon) Interglaciation, perhaps 110,000 yr B.P., and included at least two major ice advances and retreats. These glacial events took different forms in different regions. In eastern North America, glacial ice from arctic Canada came together to form the huge Laurentide Ice Sheet, which covered nearly all of eastern and central Canada and spread south to New England and beyond the Great Lakes in the Midwest. The western edge of the Laurentide Ice Sheet advanced close to the east side of the Rocky Mountains in Montana, with a lobe of ice protruding almost as far south as the location of Great Falls (Fig. 5.7). In the west, the buildup of glacial ice in the mountains of British Columbia formed the Cordilleran Ice Sheet, which butted up against the Lauren-

Figure 5.6. McDonald Valley, Glacier National Park: a U-shaped glacial valley. (Photograph by Paul Carrara, U.S. Geological Survey.)

tide Ice Sheet in western Canada and flowed south in Montana to the site of Flathead Lake. The two ice sheets covered more than 16 million km² (6 million square miles) of North America and contained a third of all the ice bound up in the world's glaciers during the last glaciation.

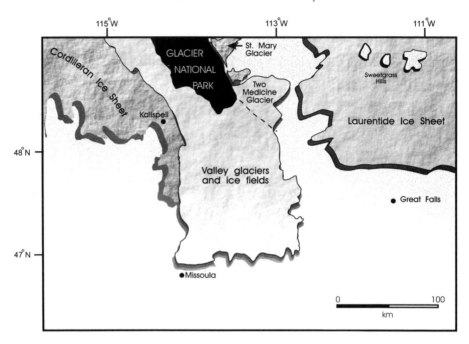

Figure 5.7. Map of northwestern Montana at the height of the Pinedale Glaciation, showing approximate boundaries of the Cordilleran and Laurentide ice sheets, valley glaciers, and ice fields. Note that high-elevation terrain in the Sweetgrass Hills protrudes above the ice. See text for explanation.

The region now known as Glacier National Park was uniquely situated to record the events of the last glaciation because it was hemmed in by the two great ice sheets. Not only that, but mountain glaciers in the park region built up enough ice to spill down several drainages onto the adjacent lowlands (Fig. 5.8).

The park is dominated by two mountain ranges trending northwest-southeast. On the west side of the park, the Livingstone Range runs about 35 km (22 miles) from the park boundary to Lake McDonald. To the east, the Lewis Range extends the full length of the park, a distance of about 100 km (61 miles) from the park boundary on the north to Marias Pass on the south. Today, many small glaciers cling to **cirque** walls near the tops of these mountain ranges. The Continental Divide runs along the crest of the Lewis Range from Marias Pass to about 16 km (10 miles) south of the U.S.-Canadian border; then it swings west to the crest of the Livingstone Range into Canada. In most of the park, the Continental Divide also served as an ice divide during the Pinedale Glaciation; western-slope glaciers flowed west from the divide and eastern-slope glaciers flowed east. The two ranges spawned glaciers that flowed tens of kilometers, carrying millions of tons of ice.

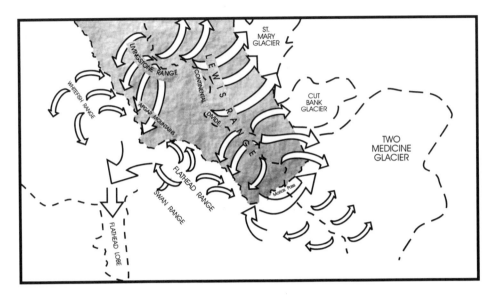

Figure 5.8. Geologist Paul Carrara's interpretation of ice flow patterns in and around Glacier National Park during the Pinedale Glaciation. Arrows indicate general ice flow directions. Note that ice crossed over the Continental Divide at Marias Pass.

At the height of the Pinedale Glaciation, about 20,000 years ago, only the highest ridges and peaks in the park were free of ice (Fig. 5.9). These ice-free regions are called **nunataks.** The remainder of the park was a sea of streaming glaciers and ice fields. These features have been mapped by my colleague, Paul Carrara. He worked out the ice flow patterns summarized in Figure 5.8 based on the position of moraines, the orientation of striations and grooves carved in bedrock by glaciers, and other features left behind by the ice. Mountain glaciers that formed on the western slope of the Continental Divide in the Livingstone Range flowed down the valley of the North Fork of the Flathead River. These glaciers joined with ice coming from the eastern slope of the Whitefish Range (west of the park boundary), and the combined glaciers flowed south, overriding the lower Apgar Mountains and contributing ice to the Flathead Lobe of the Cordilleran Ice Sheet. Glaciers from the Flathead and Swan ranges also contributed ice to this lobe. The Flathead lobe gouged out a deep depression, and the terminal moraine left by this glacial lobe dammed the Flathead River about 20,000 yr B.P., creating Flathead Lake.

Glaciers that flowed from the western side of the Lewis Range south of Lake McDonald merged with ice from the southeastern flank of the Flathead Range. This ice flow, combined with other mountain glaciers southeast of the park, formed a large body of ice around the southwestern corner of the park. This body

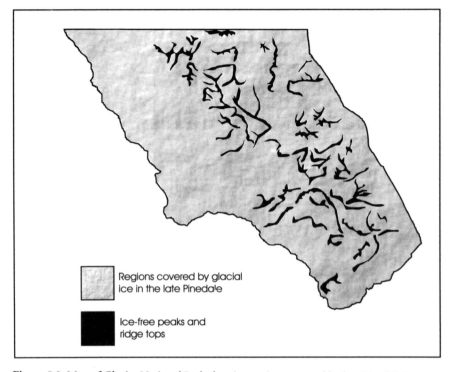

Regions covered by glacial
ice in the late Pinedale

Ice-free peaks and
ridge tops

Figure 5.9. Map of Glacier National Park showing regions covered by late Pinedale ice, and the ridge tops and peaks that rose above the ice.

pushed up and over the low divide at Marias Pass and ended up in the large **piedmont glacier** called Two Medicine Glacier. This glacier was also fed by ice flowing from the southern end of the east slope of the Lewis Range. Farther north along the eastern slope of that range, mountain glaciers flowed out onto the plains to form the Cut Bank and St. Mary glaciers. Ice from the northwestern flanks of the Lewis Range and the northeastern flank of the Livingstone Range flowed north into Canada.

The Pacific Northwest region contains a number of volcanoes that were active during the late Pleistocene and Holocene. Mount St. Helens erupted many times during this interval, spewing forth great quantities of ash. One of these ashes, labeled the Mount St. Helens "J" ash, happened to fall on Glacier National Park just as the region was emerging from the grip of Pinedale ice. Another, the Glacier Peak ash, fell about 200 years later. These ashes provide very convenient white marker horizons in sediments deposited at that time (Fig. 5.10). The Glacier Peak ash is more easily found in the Glacier National Park region, because it fell in a thicker layer than the Mount St. Helens "J" ash. Glacier Peak is a volcano located in

Figure 5.10. Exposure of lake sediments excavated at the Marias Pass site, Glacier National Park. The thin white layer near the bottom of the exposure is the Mount St. Helens "J" ash. The thicker white later above it is the Glacier Peak ash. The light gray layers above the Glacier Peak ash represent lake sediments containing Glacier Peak ash that washed in from the surrounding landscape over a period of several years after the volcanic eruption. (Photograph by Paul Carrara, U.S. Geological Survey.)

North Cascades National Park, Washington. A willow twig collected by Paul Carrara from immediately below the Mount St. Helens ash has been radiocarbon dated at about 12,000 yr B.P. This twig was too small to yield a conventional ^{14}C date, but it contained ample carbon for an accelerator mass spectrometer date. The deglaciation was a fairly rapid event, because by 11,200 yr B.P. the ice had retreated to local mountain valleys in the park, and by 11,000–10,000 yr B.P. it had withdrawn to the same high-mountain cirques and shaded niches where it remains today.

One regional phenomenon caused by glacial melting had only indirect effects on Glacier National Park, yet it was so unique and dramatic that I must pause here and summarize it. That phenomenon was the repeated filling and catastrophic draining of two large proglacial lakes. These were lakes of glacial meltwater that formed at the edges of ice sheets.

West of Glacier National Park, Glacial Lake Missoula began forming along the southern edge of the Cordilleran Ice Sheet about 15,000 yr B.P. (Fig. 5.11) when the Clark Fork River was dammed by a lobe of ice, called the Purcell Trench Lobe. This

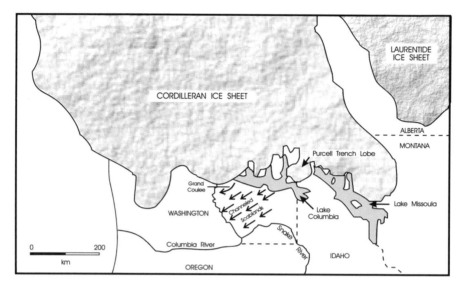

Figure 5.11. Map of the Pacific Northwest region at about 15,000 yr B.P., showing glacial lakes Missoula and Columbia, and the flow pattern of catastrophic drainage into the Columbia River Valley that created the Channeled Scablands and Grand Coulee of eastern Washington.

lake was about the size of modern Lake Ontario. The tip of that lobe occupied the basin now filled by Lake Pend Oreille, Idaho.

When Lake Missoula rose to a certain height, the ice dam of the Purcell Trench Lobe began to float. The water in the lake then burst out under the floating ice in a catastrophic flood. It has been estimated that this flood released 21 million m³ (27.5 million cubic yards) of water per second. To put this in perspective, the largest flood in recent history, a flood on the Amazon, released 385,000 m³ (503,600 cubic yards) per second.

The floodwaters streamed into Glacial Lake Columbia to the west and rapidly overwhelmed that basin. From there, the flood swept over land to the Columbia River Valley. This catastrophe is thought to have been repeated more than forty times in 1500 years! It undoubtedly wrought havoc with regional ecosystems, as it destroyed nearly everything in its path. The landscape that bore the brunt of the floodwaters still has not recovered. It is known as the Channeled Scablands of eastern Washington.

I recently flew over this region, and at a height of 30,000 ft enormous ripple marks remain clearly visible. The largest runoff channel is called Grand Coulee. Here, the floodwaters carved a deep trench into the bedrock. This is but one of many regions in North America that owe their modern topography to events that took place in the late Pleistocene.

After the Ice Age: The Return of Tall Trees and Alpine Meadows

As we've seen, the Glacier National Park region was just about completely locked in ice during the last glaciation. An ecosystem resembling the **Beringian steppe-tundra** bordered the southern edge of the ice sheets and local glaciers. It was essentially a cold prairie, with a mixture of arctic and grassland species. For instance, southwestern Montana and northeastern Wyoming were home to mammoths, musk-oxen, and caribou. On the other hand, these regions also supported populations of prairie dogs, sage voles, and pocket gophers. None of the latter group of animals is adapted to arctic conditions, so the climate, although decidedly colder than today, was not an arctic climate. Fossil insects from Pinedale Glaciation times that I have described from Lamb Spring, near Denver, are indicative of the type of cold grassland environments found today at the northern edge of the prairie in Alberta. So the biota that lived south of the ice sheets in the Rocky Mountain region was a curious mixture of arctic and prairie species, one in which mammoths and caribou foraged for grasses in the company of prairie dogs. Winter snows persisted on the ground longer than they do now, and the summers were cool.

Based on the fossil insect data, summer temperatures on the plains east of the Colorado Front Range at 15,000 yr B.P. were about 10°C colder than they are now. This cooling estimate agrees with that derived from other lines of evidence, including estimates of the type of climate required to grow the regional glaciers to their Pinedale size. As I've mentioned before, winter temperatures need not have been much cooler than they are today in order for the glaciers to have grown. Even our modern winters in the Rockies are generally cold enough to keep snow from melting until summer. It is the summer temperatures that matter most.

Thus the cold conditions alone may account for the growth of Pinedale glaciers. Some paleoclimate models suggest that there was actually less precipitation in late Pinedale times than there is today. The glaciers built up because the snow that fell did not melt from one year to the next. However, it is important to stress that glacial dynamics are complex, with many conditions feeding back on one another. For instance, once Glacial Lake Missoula formed from melting Cordilleran ice, it became an added source of moisture for precipitation feeding the mountain glaciers in northwestern Montana.

How did the flora and fauna of the northern Rockies respond to the Pinedale deglaciation? The evidence is fragmented, but several fossil localities in northwestern Montana provide bits and pieces of the regional story. To the east of Glacier National Park, palynologist Cathy Whitlock of Oregon State University has studied the transition from late glacial to Holocene environments from pollen in the sediments in Guardipee Lake and the history of the Holocene prairies at Lost Lake

(Fig. 5.2). West of the park, Richard Mack and colleagues analyzed postglacial pollen from Tepee Lake and McKillop Creek Pond in the Kootenai Valley. At Marias Pass, just south of the park, my colleague Susan Short studied pollen in sediments from a late glacial pond, and I analyzed the insect fossils from those sediments.

The Guardipee Lake basin was covered by the Two Medicine Glacier during late Pinedale times. The glacier retreated and sediments started accumulating just before Mount St. Helens "J" ash was deposited in the lake, about 12,000 yr B.P. The pollen from this lake provides a record of regional vegetation that begins just after deglaciation (Fig. 5.12A). The vegetation cover that developed on the newly exposed landscape was a grassland with some sagebrush. Nearby mountain slopes to the west were already supporting pines, spruce, and fir. Closer to the lake, junipers were becoming established. Juniper apparently does not compete well with other conifers in regions where they are all capable of growing, but in the early postglacial times tree species were only just beginning to disperse across vast deglaciated landscapes, and junipers were one of the first conifer groups to take advantage of these landscapes. Willow and aspen were probably also growing in local wet areas during deglaciation.

Even though regional climates to the south were warming up rapidly by 12,000 yr B.P., sites close to the waning ice margins remained somewhat cooler because of the chilling effects of the ice and the large volumes of near-freezing meltwater accumulating in glacial lakes. The insect assemblage from Marias Pass also dates to about 12,000 yr B.P. It is an interesting mixture of arctic, alpine, and boreal beetles and ants. Taken together, the insect assemblage indicates that treeline, the altitudinal limit of trees, was just below the fossil site. This finding suggests that the conifer forest had gotten a foothold on the lower slopes of the mountains but had not advanced upslope to the fossil site's elevation (about 1500 m [4920 ft]). Above that, tundra, **fellfields,** and the remaining ice were holding on.

The climatic conditions necessary to keep treeline depressed to below 1500 m translate to temperatures about 5°C cooler than they are today. In some studies of slightly younger insect fossils from high mountain sites, I have drawn different conclusions from the faunas. Younger faunas from the Colorado Rockies (dating to about 10,000 yr B.P.) contain insects found at or near the study sites today, indicating that the climate was already as warm as it is today, even though the conifer forests had not yet climbed to that elevation. But in the case of the Marias Pass insects, no species in the fossil assemblage can be found at that elevation today. All of the species are found either much farther north, in the subarctic and arctic regions, or at much higher elevations in the mountains. Therefore I have concluded that summer temperatures at Marias Pass 12,000 years ago were not as cold as those during the height of the Pinedale Glaciation, but also not as warm as

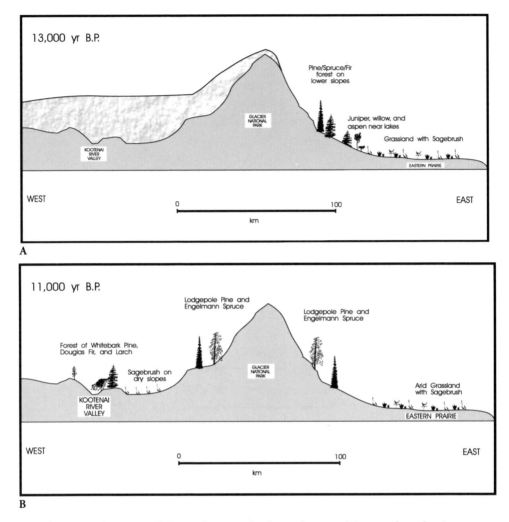

Figure 5.12. Summary of changes in vegetation in northwestern Montana from the time of deglaciation through the late Holocene, based on data from the Kootenai River Valley, Marias Pass, and various bogs in Glacier National Park, and Guardipee and Lost lakes on the eastern prairie. Summaries of regional vegetation are shown for 13,000 yr B.P. (A), 11,000 yr B.P. (B), 9500 yr B.P. (C), 6000 yr B.P. (D), and 2500 yr B.P. (E).

those today. Susan Short also reached this conclusion, based on a study of the pollen from the Marias Pass lake sediments of the same age.

By 11,500 yr B.P., the vegetation surrounding Guardipee Lake reflects a change in climate. An increase in aridity was inferred from a rise in sagebrush and other dry-land plant pollen, at the expense of grasses and other herbs. In addition, plants

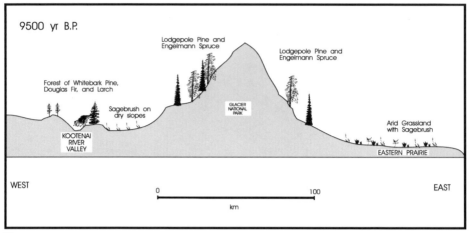

C

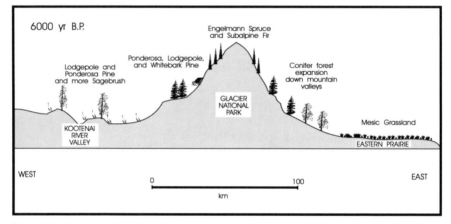

D

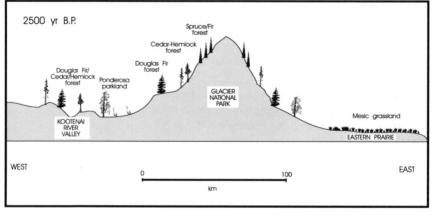

E

Figure 5.12. (*continued*)

adapted to brackish water started growing in Guardipee Lake. This finding suggests that the lake level was dropping. The lake dried out after 9300 yr B.P. and then went through a series of wet and dry episodes for the rest of the Holocene.

Meanwhile, in the Kootenai River Valley west of the park, the sediments in Tepee Lake began accumulating and collecting pollen. The Cordilleran Ice Sheet withdrew from the site before 11,200 yr B.P. (the age of the Glacier Peak ash). By this time, regional climate was clearly quite warm, because the newly deglaciated terrain was immediately colonized by pines (probably whitebark pine, based on the shape and size of the pollen grains). Smaller numbers of spruces were also present, with a ground cover of grasses (Fig. 5.12B). By about 11,000 yr B.P., conifer forest was growing right up to the edge of the waning Cordilleran Ice Sheet.

From 11,000 to 7000 yr B.P., regional forests were dominated by pines, Douglas-fir, and/or larch. Open ground regions, such as meadows and dry hillsides, were covered with sagebrush. Based on the pollen evidence, the climate during this interval was warmer than the climate during the Pinedale deglaciation.

In Glacier National Park, Paul Carrara has obtained conifer wood from a number of peat bogs by taking cores through the peat. The wood has been radiocarbon dated and identified (based on the cell structure of the wood), and it provides the following scenario for the reforestation of the west side of the park after the ice retreated. Spruce or larch trees were growing in the Lake McDonald region by 10,600 yr B.P., and lodgepole pine was there by about 10,000 yr B.P. Forests climbed to an elevation of at least 1350 m (4430 ft) by 9600 yr B.P., based on wood from a bog on Snyder Ridge, above the lake (Fig. 5.12C).

In the Kootenai River Valley, drier conditions prevailed after 7000 yr B.P., and dry-adapted conifers (ponderosa and lodgepole pine) became regionally important. This aridity was especially strong at about 6700 yr B.P., when sagebrush replaced conifers around Tepee Lake and nearby McKillop Lake (Fig. 5.12D). After 4000 yr B.P., moister conditions returned, accompanied by Engelmann spruce. Modern climatic conditions and vegetation became established about 2700 yr B.P., heralded by the establishment of western hemlock, a tree that characterizes modern forests from western Washington to northwestern Montana (Fig. 5.12E).

On the plains east of the park, the pollen record from sediments in Lost Lake records changes in prairie vegetation through the Holocene. From 9400 to about 6000 yr B.P., this region was an arid grassland with a drier climate than is seen today (Fig. 5.12C). Following 6000 yr B.P., cooler, moister conditions allowed a more mesic grassland to develop on the plains, shrubs increased in wetland regions, and conifer forest expanded in nearby mountains (Fig. 5.12D).

Based on the evidence from the Kootenai River Valley, the modern ecosystems of northwestern Montana did not become established in their current form until 8000 years after Pinedale ice retreated from this region. This finding shows us that regional ecosystems are not very steady, long-lived associations of plants and animals. Rather, they are dynamic. Who knows how long the current set of creatures in northwestern Montana will remain in the same associations? They appear stable to our eyes, mostly because human beings have such short lives in comparison with ecosystems. With the exception of such regional disturbances as the eruption of Mount St. Helens in 1980, no ecologist has ever seen an entire ecosystem change dramatically, except on account of disturbances brought about by fellow humans.

The other important point to take away from this evidence is that ecosystem composition is fine-tuned to such environmental limits as slight changes in annual precipitation or the time of year when that precipitation falls. Subtle changes in environmental conditions can bring about wholesale changes in forest composition. Of course, the other major dynamic in the system is the interaction between species. Competition, predation, parasitism, and other associations between members of a biological community can force one species to die out and allow another species to get a foothold in a region. It is said that nature abhors a vacuum. An empty niche in an ecosystem does not stay empty for long.

Glacial Advances during the "Little Ice Age"

The period between about A.D. 1500 and 1900 has been called the "Little Ice Age" by historians and climatologists because much of the Northern Hemisphere experienced temperatures far colder than average. The cooling brought on a buildup of glacial ice in the mountains and in the arctic. Pack ice clogged sea lanes around Iceland and Greenland. Glaciers in the Alps advanced dramatically, overrunning alpine meadows. In the northern Rocky Mountains, glaciers advanced out of their cirques and spilled down mountainsides, forming large moraines. These moraines are clearly visible today in front of many of the remaining glaciers in Glacier National Park. Based on tree-ring dating, the glaciers in the park reached their Holocene maximum during the mid-nineteenth century.

At the Jackson Glacier, Little Ice Age moraines (also called Neoglacial moraines by geologists) are well marked (Fig. 5.13). Two ages of moraines are visible in front of Red Eagle Glacier (Fig. 5.14). The steep, unvegetated moraines closest to the glacier were formed during the Little Ice Age. The lower, vegetated moraines downslope from the Little Ice Age moraines were probably formed at the end of the Pinedale Glaciation.

As I mentioned in the introduction to this chapter, the park may owe its existence to this fact, because the nineteenth-century visitors to the region were so impressed by the glaciers that they lobbied for the creation of the park. Were they to see the park today for the first time, they might still come away impressed by the

Figure 5.13. Mount Jackson and the Jackson Glacier, Glacier National Park. The 1850 moraine can be seen in the foreground. (Photograph by Paul Carrara, U.S. Geological Survey.)

Figure 5.14. The Red Eagle Glacier region, showing two ages of moraines. The steep, rubbly, unvegetated moraines (closer to the modern ice limit) are Little Ice Age moraines; the lower, subdued, vegetated moraines farther downslope were probably left by the retreating ice margin of a Pinedale glacier, about 10,000–11,000 yr B.P. (Photograph by Paul Carrara, U.S. Geological Survey.)

beauty and grandeur of the landscape, but the word *glacier* might not feature so prominently in their descriptions. Nevertheless, the park landscape owes so many of its features to the action of glaciers that the name Glacier National Park will always be appropriate, even if the current crop of glaciers melt completely away. After all, it is extremely likely, if not inevitable, that world climate will cool off once again and another ice age will seize this region in its frigid grasp.

Since the mid-nineteenth century, the glaciers in the park have shrunk drastically. At the start of this decline, there were more than 150 active glaciers in the park. Between then and the beginning of the twentieth century, the glaciers still covered about the same area, but they decreased in thickness. By 1940, the glaciers had retreated greatly, and many had melted away entirely. According to geologist Mark Meier, only 53 glaciers were left by 1958. Even the largest glaciers of the nineteenth century just barely exist today.

The Agassiz Glacier was once one of the largest ice masses in the park, covering several square kilometers in the mid-nineteenth century. It is estimated to have been 200 m (650 ft) thick at that time. The area of the ice from that advance can still be seen below the glacier, where the trees were eliminated by the ice. The boundary of this poorly vegetated region is called the glacial *trimline*. Inside it, trees are either absent or substantially younger than in the surrounding forests.

Figure 5.15. The Agassiz Glacier region, showing remaining snowbank and forest trimline dating from the mid-nineteenth century, when this glacier stood at its Holocene maximum. (Photograph by Paul Carrara, U.S. Geological Survey.)

The Agassiz Glacier is now in danger of losing its "active glacier" status. The ice has shrunk to about 10 or 15 m (33–50 ft) in thickness, and it now represents less than 3% of its former volume (Fig. 5.15). When an ice mass stops flowing downhill, it is no longer called a glacier.

Modern Ecosystems of the Park

The modern ecosystems in the park region fall into six altitudinal zones (Fig. 5.16). The lowland regions west of the park remain semiarid grassland or sagebrush steppe, dominated by sagebrush, bunch grass, and western wheat grass. The Great Plains grassland east of the park is a short grass prairie, with western wheat grass, buffalo grass, and needle grass. The montane, subalpine, and alpine vegetation zones of the park differ sharply between eastern and western slopes. The increased Pacific moisture on the western slope allows a forest of ponderosa pine and Douglas-fir to grow at elevations between about 900 and 1200 m (2950–3940 ft),

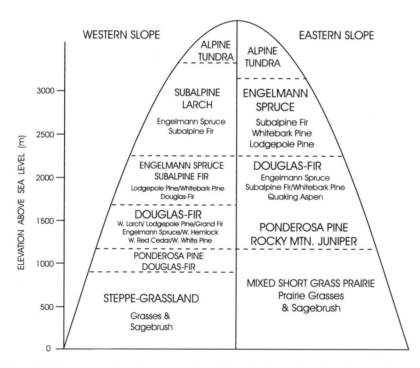

Figure 5.16. Summary of modern vegetation of the Glacier National Park region. (After Barnowsky et al., 1987.)

while grassland persists on the eastern slope to an elevation of 1200 m. On the western slope, Douglas-fir forests dominate slopes from about 1200 to 1600 m (3940–5250 ft). These forests contain abundant elements of Pacific Northwestern forest trees, including western larch, grand fir, western hemlock, western red cedar, and western white pine.

In this elevational range on the eastern slope, a ponderosa pine–Rocky Mountain juniper forest is dominant, giving way to Douglas-fir–dominated forests above about 2000 m (6560 ft). These forests also contain Engelmann spruce, subalpine fir, whitebark pine, and quaking aspen.

Between about 1750 and 2200 m (5740–7220 ft), western slope forests are dominated by Engelmann spruce and subalpine fir, although lodgepole pine, whitebark pine, and Douglas-fir also grow there. The forests of the western slope regions in the park would probably be better defined in well-established altitudinal zones if they had remained undisturbed by fire, disease, and other vicissitudes during the past few centuries. As it is, many species have gained footholds above or below their expected elevation ranges because of ecological disturbances.

Subalpine larch dominates the upper subalpine regions of the western slope of the park, from about 2250 to 3200 m (7380–10,500 ft). Engelmann spruce and subalpine fir are also found there. On the eastern slope, Engelmann spruce dominates the subalpine zone, along with subalpine fir, whitebark pine, and lodgepole pine. Upper treeline on the eastern slope is about 200 m (655 ft) lower than on the western slope. Many of the same tree species occur on both eastern and western slopes, but in different communities at different elevations. In addition, the western slope forests contain many Pacific Northwestern elements that are lacking on the drier eastern slope, where the forests are more similar in composition to those found farther south along the Rocky Mountain chain.

The Park's Human History

Brian Reeves, an archaeologist at the University of Calgary, Alberta, has been studying the history of human activity in the Glacier National Park region. The park has been out of American Indian hands since the Blackfoot Tribe ceded the eastern edge of the park to the U.S. government at the beginning of the twentieth century. Prior to that, native peoples had been camping and hunting in the park region throughout the Holocene. The alpine zone (above treeline) was apparently an important hunting ground in prehistory, as 220 sites that span the last 10,000 years have been discovered there. High mountain passes, such as Boulder Pass, were used by prehistoric peoples as they are by modern-day hikers and campers. In fact, many campsites favored by modern park visitors were also used repeatedly by Indians. This should come as no surprise. A good campsite is a good campsite, whether you are pitching an ancient tent or a modern one. Typically, these are sites with unobstructed views to neighboring valleys and easy access to water and wood. Modern-day campers need to be aware of this history and to report any accidental archaeological discoveries, such as stone tools, to a park ranger.

At Boulder Pass, prehistoric hunters camped near a Neoglacial (late Holocene-age) moraine and hunted bighorn sheep using game drives. Game drives generally consisted of two low stone walls, converging over a few hundred meters to funnel game animals to an ambush point. The highlands of Many Glacier Valley contain several game drive walls, as well as cairns, which were piles of rock left as route markers. Brown's Pass also contains evidence of prehistoric hunting camps. In addition to hunting, ancient peoples also developed small quarries in exposed seams of workable rock, from which they dug stones from which to make tools.

During the last 1000 years, use of the alpine zone apparently tapered off, possibly because of climatic deterioration. Cooler, wetter conditions led to more permanent snowbeds at high elevations, keeping both game animals and hunters off the high terrain.

In St. Mary Valley (Fig. 5.17), an extensive prehistoric base camp has been found along the northwest lakeshore. This site was occupied for as long as 8000 years,

Figure 5.17. St. Mary Lake, Glacier National Park. The shores of this lake were used repeatedly as campsites by prehistoric peoples. (Photograph by Paul Carrara, U.S. Geological Survey.)

apparently during the fall and spring seasons. Fishing tools have been found there, including sinker stones for nets. Bison and bighorn sheep were also hunted in the area. Many prehistoric peoples used lakeshores for bison hunting. The animals coming to the lake to drink were driven out into the water to be killed. A bison floundering around in the water is a much easier target for a spear or arrow than a bison charging across the plains at 40 miles per hour!

Indians occupied the whole Glacier National Park region throughout prehistory. They were drawn, just as we are, to the high mountains and lakes. So far, forty Indian vision quest sites have been found in the park. These are high mountain localities, such as East Flattop Mountain and Squaw Mountain, where men and boys went to fast, pray, and seek visions from their gods.

Glacial Features Easily Viewed in the Park

Although the glaciers themselves are disappearing, they have left some spectacular scenery behind. Some of the glacial features are illustrated in Figure 5.18. The deep, steep-walled valleys of the park are due to glacial scouring. The sheer cliffs and sharp peaks in the park also owe their shape to sculpting by glacial ice. As ice flowed over bedrock, it eroded surfaces to different degrees, creating benches and stairstep valleys. The long, sharp mountain ridges in the park, including Garden Wall and Pinnacle Wall, are known as *arêtes*. They were formed when ice scraped

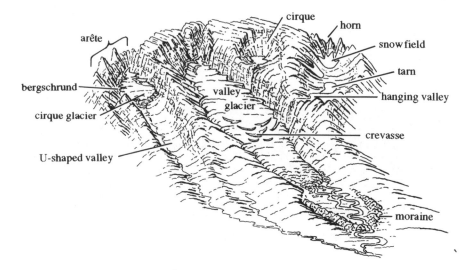

Figure 5.18. Mountain glacial features prominent in Glacier National Park. (From Carrara, 1993; sketch by T. R. Alpha, 1991.)

Figure 5.19. Flinsch Peak, Glacier National Park: a glacial horn. Oldman Lake (foreground) is a tarn. (Photograph by Paul Carrara, U.S. Geological Survey.)

along both sides of a ridge, leaving over-steepened slopes that come to a sharp crest at the top (Fig. 5.18).

The high, steep-walled valleys perched near the tops of many mountains in the park are *glacial cirques*. Some remain filled with glacial ice. The boundary between flowing glacial ice and a snowfield at the glacier's edge is called a *bergschrund*. This zone of a glacier may develop one or more deep crevasses. In many cirques in the park, the glacier has melted, leaving the cirque walls exposed. The depression at the base of the cirque is often filled with water, forming a cirque lake or *tarn*. Where a larger valley was eroded by a large glacier and a smaller, higher valley was eroded by a tributary glacier, *hanging valleys* form. These glacial valleys have a mouth that is perched at a relatively high level on the steep side of a larger glacial valley.

Pyramid-shaped peaks called *horns* were shaped by glaciers grinding their way along three or more sides of a mountain (Fig. 5.19). Examples of these "American Matterhorns" include Flinsch Peak, Fusillade Mountain, Reynolds Mountain, and Thunderbird Peak. Many of the park's most beautiful waterfalls originate in cirques and hanging valleys carved by glacial ice. Some of the falls drop into streams still choked with a heavy load of glacial debris. These streams, called *braided streams,* are constantly shifting across numerous channels as they make their way through valleys filled with glacial sediments.

The glacial ice in the park was more spectacular during the last century than it is today, but the effects of glaciation are everywhere to be seen. Ice shaped the park's features and made it more attractive to the human eye. On the geologic time scale, Glacier National Park has only just been released from its mantle of ice. Its peaks have been freshly carved, its valleys freshly filled with debris. Here the Holocene may be looked on as a brief respite from the landscapes' domination by ice and snow. Did we not happen to live in such an unusually warm era, our chances on a Pleistocene time scale of seeing Glacier National Park in such relatively ice-free condition would be only about one in twenty.

Suggested Reading

Alt, D. 1983. Geology: Glaciated park. In Alt, D., Buchholtz, C. W., Gildart, B., and Frauson, B. (eds.), *Glacier Country: Montana's Glacier National Park*. Helena: Montana Magazine, pp. 6–25.

Barnowsky, C. W., Anderson, P. M., and Bartlein, P. J. 1987. The northwestern U.S. during deglaciation: Vegetational history and paleoclimatic implications. In Ruddiman, W. F., and H. E. Wright, Jr. (eds.), *Geology of North America*, Volume K-3: *North America and Adjacent Oceans during the Last Deglaciation*. Boulder, Colorado: Geological Society of America, pp. 289–321.

Carrara, P. E. 1989. Late Quaternary Glacial and Vegetative History of the Glacier National Park Region, Montana. U.S. Geological Survey Bulletin No. 1902. 64 pp.

Carrara, P. E. 1993. Glaciers and Glaciation in Glacier National Park, Montana. U.S. Geological Survey Open File Report 93-510. 18 pp.

Harris, A. G. 1977. Glacier National Park. In *Geology of the National Parks*. 2d ed. Dubuque, Iowa: Kendall/Hunt, pp. 65–76.

Mack, R. N., Rutter, N. W., and Valastro, S. 1983. Holocene vegetational history of the Kootenai River Valley, Montana. *Quaternary Research* 20:177–193.

6

YELLOWSTONE NATIONAL PARK
A Land of Fire, Ice, and Water

Yellowstone National Park is often called the jewel in the crown of our national parks system. It was the first national park to be established in the United States (in 1872), and it was also the world's first national park. Yellowstone is perched on a series of high volcanic plateaus in the northwest corner of Wyoming (Fig. 6.1). The average elevation of the plateau regions is about 2400 m (7875 ft) above sea level. Ringed by mountains that climb to more than 3300 m (10,830 ft) elevation (Fig. 6.2), the park receives substantial moisture (but much less than Glacier National Park), especially in winter. The park's elevation and northern latitude make for a cold climate. The average yearly temperature on the Yellowstone Plateau is less than 1°C (33°F), and snow covers the ground for half the year. The climate is cold enough to sustain patches of permafrost. Permafrost occurs at many high-elevation sites but also exists at elevations as low as 2100 m (6890 ft) on a north-facing slope near Tower Junction.

The precipitation pattern in the park is tied strongly to elevation and mountain topography. The north entrance to the park, near Gardiner, Montana, receives only 200 mm (8 in.) of moisture annually, whereas the elevated southwest corner of the park receives more than 1750 mm (70 in.) of precipitation.

Unlike Glacier National Park, Yellowstone's mountains contain no cirque glaciers today, although semipermanent snow lies in patches on some high, north-

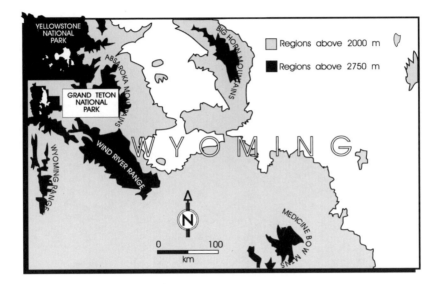

Figure 6.1. Map of Wyoming, showing the location of Yellowstone National Park, major mountain ranges, and high-elevation regions.

facing slopes. Water in the liquid state is what makes Yellowstone such a glistening jewel. Especially for the visitor who has traveled across hundreds of kilometers of sagebrush prairie to get there, Yellowstone appears as an oasis of tall forests and water. Water roars over the great falls of the Yellowstone River, rushes madly in rocky streams, and meanders gracefully in broad rivers. It fills one enormous lake and many smaller lakes, ponds, and bogs. But water's most famous appearance in the park is its emanation, boiling hot, from a fiery cauldron hidden beneath the surface. The geysers, hot springs, mud pots, and fumaroles in Yellowstone constitute the largest collection of geothermal (earth-heated) features in the world. This fact is well known. What is less well known is that, from a geologic perspective, these fascinating features are a recent phenomenon that owe their origin to events in the Pleistocene.

Modern Setting

The modern ecosystems of the park are shown in Figure 6.3. Sagebrush steppe dominates the lowlands up to an elevation of about 1500 m (4920 ft). This vegetation is especially prominent at the north end of the park, where elevations

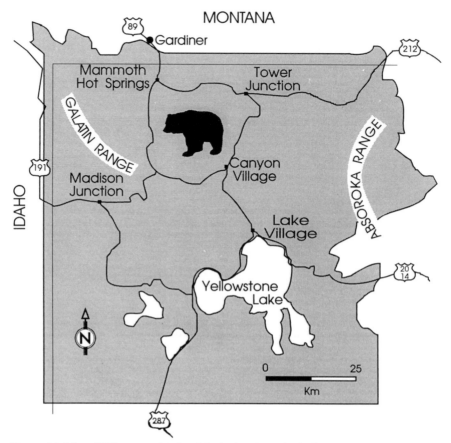

Figure 6.2. Map of Yellowstone National Park, showing principal mountain ranges.

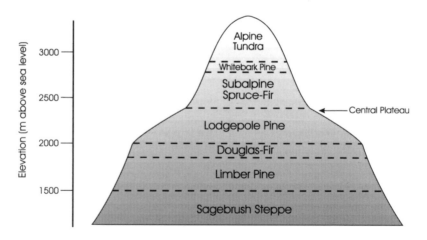

Figure 6.3. The modern ecosystems of Yellowstone National Park, shown along an elevational gradient.

are generally lower. Pine **parkland** ranges from 1500 to 2400 m (4920–7875 ft) in elevation. In the lower elevations of this range, limber pine is the dominant tree. This gives way to Douglas-fir in a narrow elevational band between about 1900 and 2000 m (6230–6560 ft), then lodgepole pine dominates up to the boundary with subalpine forest at 2400 m (7875 ft). Lodgepole pine completely dominates the central plateau region of the park. As I will discuss later, this distribution is due to the nature of the volcanic soils of the plateau. Subalpine forests in the park include Engelmann spruce, subalpine fir, and whitebark pine. Whitebark pine forms the upper treeline at about 2800 m (9190 ft). Above this elevation is the alpine tundra zone, with low-growing shrubs, herbs, and sedges.

Yellowstone National Park is home to many animal species and is famous for its bears, both black bear and grizzly. They aren't seen as often as they used to be, and this is no accident. The National Park Service has taken steps to remove the bears from roadsides and campgrounds, where they may cause trouble as they abandon their natural foods for human handouts. They now roam mostly in the back country, although grizzlies are still seen in the Hayden Valley on a regular basis. But Yellowstone is home to more than bears. Now that the gray wolf has reentered the park, Yellowstone is once again home to its full complement of wildlife. The park now contains all the elements of the flora and fauna that lived there when Europeans first visited, back in the early nineteenth century. The native fauna includes grazers and browsers, such as bison, pronghorn antelope, moose, mule deer, and elk. In addition to the bears, the fauna also includes such predators as the mountain lion, lynx and bobcat, coyote, pine marten, and long-tailed weasel.

Origins of the Geothermal Wonders

Standing next to a boiling hot spring or erupting geyser, it takes little imagination to realize that Yellowstone is a geologic "hot spot." Most geologic features change so slowly that they appear unchanging to human eyes. In contrast, the geothermal features of Yellowstone are very dynamic. Every few years, new hot springs or fumaroles appear, sometimes triggered by local earthquakes. Other hot springs dry up, and the frequency and size of geyser eruptions change. There are few places on the planet where the Earth is so energetic. Many a visitor to the park has wondered, "What's going on here?"

In order to understand what happens in the geyser basins at Yellowstone, we need to delve deep beneath the Earth's surface. Our planet has an outer crust of rock. This crust averages about 50 km (31 miles) in thickness over the continents and is about 16 km (10 miles) thick under the oceans. This crust serves as a layer of insulation, protecting the surface from the enormous heat of the interior or *mantle.* The mantle layer is thousands of kilometers thick. It varies in density, but in some small regions it is a fiery liquid of molten rock, called *magma.* For reasons that geologists are still struggling to understand, magma builds up in pockets beneath the crust in some regions. Where these buildups occur, the magma pools are like blisters beneath the skin, which push up and thin the stretched crust. The heat of the magma melts the lower layers of crust rock, thinning the solid crust layer even further.

Yellowstone sits over one of these magma chambers. As a result, the Earth's crust in the Yellowstone region is only about 5 km (3.1 miles) thick. In other words, the crust is 90% thinner here than in most continental regions. As might be expected, a crust under this kind of stress is riddled with cracks as it is stretched and thinned over a growing cauldron of magma near the surface. Water from rain and meltwater from snow seep down into these cracks, where they become superheated. As the heated water builds up in the cracks, it expands and is forced up and out at the surface. Depending on the shape and size of the "plumbing" beneath the earth, that water escapes as steam (in fumaroles), as boiling pools of water (in hot springs), or in a mixture of steam and water under tremendous pressure (as geysers). Geysers erupt in fits and spurts because each eruption releases part of the pressure in their "plumbing," and it takes time for that pressure to build again.

There are several variations on the geothermal theme. For instance, mud pots are acidic hot springs that hold large amounts of minerals in suspension beneath the surface but do not contain enough water to flush away the material they bring to the surface. Bubbles in the muddy waters push to the surface, creating the familiar *glop-glop-glop* sound in the mud pot. Mud volcanoes release their subterranean gases in more explosive bursts.

Some hot springs, such as Mammoth Hot Springs, percolate up through bedrock that contains minerals that are readily dissolved, notably the calcium carbonate in limestone. The calcium carbonate dissolves in the hot water beneath the surface then precipitates out of the water as it cools when it reaches the surface, forming terraces of fragile *travertine.*

I have mentioned that these features were created in the Pleistocene. The geologic events that preceded and then spawned the geothermal features were so large and devastating as to beggar description. It is no exaggeration to say that about 600,000 years ago, much of the region we know as Yellowstone blew up and then collapsed into a pool of molten rock. If this same event took place today, the shock wave would be felt by everyone in the world. This is what happened. During the early Pleistocene (more than one million years ago), pressure began building up in the magma chamber beneath Yellowstone. This buildup forced the crust up and resulted in several eruptions of lava through cracks in the crust. Eventually, the pressure in the magma chamber built to the point of exploding. The explosion emptied the magma chamber, leaving the thin crust above unsupported. That roof

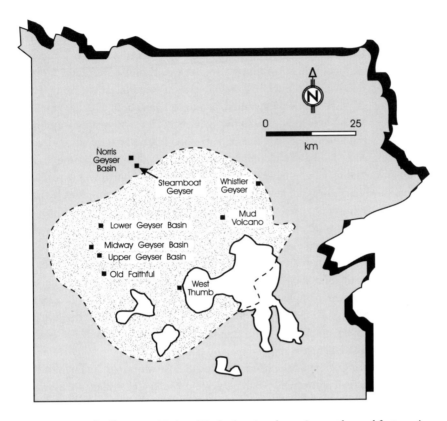

Figure 6.4. Map of Yellowstone National Park, showing the major geothermal features in relation to the boundary of the caldera region.

of crust collapsed into the empty chamber below, forming an enormous basin or *caldera* (Fig. 6.4).

Since that momentous occasion, lava has continued to ooze spasmodically out of cracks in the caldera, gradually filling in much of the depression and spilling out over the edge in some places. The upper and lower falls on the Yellowstone River spill over ledges created by two such lava flows. The original walls of the caldera are still visible in some places. The north wall of the Madison River canyon is one example.

Glacial History

The two most recent large-scale glaciations of the Rocky Mountain region are called the Bull Lake and Pinedale glaciations. Both are named after localities in the Wind River Range of Wyoming. Estimates of the age of the Bull Lake Glaciation have varied widely over the last 30 years of Quaternary research. Initially, Bull Lake was thought to have preceded the Sangamon (last) interglacial period. If this is so, then the Bull Lake Glaciation ended before about 130,000 years ago. During the 1960s, the general consensus among geologists was that the Bull Lake advances were the first ice advances following the Sangamon interglacial warming. In this scenario, the Bull Lake Glaciation would be younger than about 110,000 yr B.P. The Yellowstone region has contributed to the age debate in an interesting way. Since it is a volcanically active region, occasional lava flows in Yellowstone coincided with ice advances. When molten lava cools rapidly, its atoms do not have time to orient themselves in the usual crystalline structure. Rather, they form the natural glass *obsidian*. Obsidian can be dated by a chemical method called obsidian hydration and by the radiometric method of potassium/argon (K/Ar) dating. Bull Lake moraines underlie (and hence are older than) a basaltic flow in West Yellowstone that was dated by K/Ar at 117,000 years old. The rhyolite (extruded, fine-grained volcanic rock) flows that formed Obsidian Cliffs have been dated by obsidian hydration at about 150,000 years old. The obsidian formation at Obsidian Cliffs is thought to have been brought about when Bull Lake ice met molten lava. So Bull Lake ice appears to have formed in the Yellowstone region before the last interglacial, probably about 150,000 years ago, according to a paper by U.S. Geological Survey geologist Ken Pierce.

In most regions of the Rocky Mountains, Bull Lake moraines are farther down-valley than Pinedale moraines. Bull Lake ice advances extended an average of 20 km (12.4 miles) farther than subsequent Pinedale Glaciation advances in the western Yellowstone region. But on the north side of the park, Pinedale ice pushed out beyond the Bull Lake limit, obliterating the terminal moraines left by the older glaciers. In some regions, the glaciation that preceded the Pinedale appears to have taken place after the Sangamon Interglacial. In other regions, it appears to have preceded it. So the term "Bull Lake" is not clearly defined as a glacial event in a single interval of time.

The Pinedale Glaciation brought an immense cover of ice to the Yellowstone region. The Yellowstone Ice Cap was centered along a north-south axis through Yellowstone Lake, with ice flowing radially to the northeast, west, and southwest, in a line about 150 km (93 miles) long. Mountain glaciers from the Absaroka and Galatin ranges and Beartooth Highlands in the north came together to fill the Lamar and Yellowstone valleys, then flowed northwest into Montana (Fig. 6.5), converging near the town of Gardiner to form an outlet glacier that drained ice from the northern part of Yellowstone. Glaciers in the southern Absaroka Range flowed west into Yellowstone, occupying the depression now filled by Yellowstone Lake. From there, the ice flowed down the Hayden Valley. The two ice masses came together and spread out to cover all but the southwestern edge of the park region.

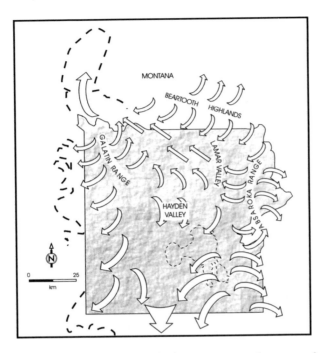

Figure 6.5. Map of Yellowstone National Park, showing extent and patterns of movement of late Pinedale glacier ice. (Data from Pierce, 1979.)

Geologists Lionel Jackson and John Clague of the Geological Survey of Canada have developed a model for the development of the Cordilleran Ice Sheet by the merging of valley glaciers (Fig. 6.6). This large-scale model essentially illustrates what happened as the Yellowstone Ice Cap developed during Pinedale times. Unlike the high peaks and ridges in Glacier National Park, the high country in Yellowstone was completely overridden by ice. The valleys and central plateau region were buried under about 700 m (2300 ft) of ice.

A recent paper by Neil Sturchio and others presents a chronology of events during the Pinedale Glaciation, based on uranium-series ages of travertine deposits in the northern Yellowstone region, particularly in and around Mammoth Hot Springs. Some of these travertine deposits are covered by Pinedale glacial deposits, whereas some others are not; some overlie older glacial deposits, and some are within preexisting glacial deposits. The relationships between the travertine and glacial deposits helped these geologists to unravel the timing of Pinedale events in the park. Between 47,000 and 34,000 yr B.P., an early Pinedale glacial advance occurred. This was followed by an **interstadial** interval, when climate warmed enough for glaciers to recede, but the level of warming did not achieve full

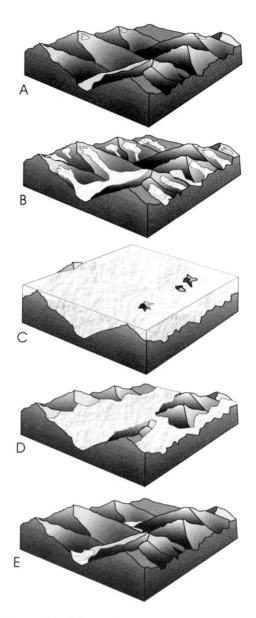

Figure 6.6. Model for the growth and decay of a regional ice cap in a mountain region. (A) During Sangamon Interglacial time, glaciers existed only in high mountain cirques. (B) At the onset of the Pinedale Glaciation, mountain glaciers grew and began flowing into adjacent valleys. (C) At the height of the glaciation, glaciers merged to form an ice cap that buried all but the highest mountaintops. (D) As deglaciation began, the ice margins receded from the high mountains but valleys remained buried in ice. (E) Near the end of deglaciation, stagnant ice masses were confined to some lowlands. (After Jackson and Clague, 1991.)

interglacial conditions. This interstadial apparently lasted from 34,000 to 30,000 yr
B.P. An extensive Pinedale ice advance took place between 30,000 and 22,500 yr
B.P., followed by a major ice recession (22,500–19,500 yr B.P.) and a minor re-
advance (19,500–15,500 yr B.P.). Ice margins retreated rapidly in most regions of
the park following 15,000 yr B.P., and park lowlands were probably completely free
of ice by about 14,000 yr B.P.

Evidence of the Pinedale Glaciation abounds in Yellowstone. A **glacial erratic**
boulder weighing about 500 tons sits near Inspiration Point, where it was deposited by a glacier. Glacial deposits (characterized by their poorly sorted load of
boulders, cobbles, sands, and silts) can be seen in Soda Creek Butte, the Lamar
Valley, and Yellowstone Canyon. Between Geode and Oxbow creeks, west of Tower
Junction, is an ancient stream deposit that was formed when meltwater ran along
the edge of a Pinedale glacier. Kettle ponds were formed in many park locations
when blocks of ice were stranded during the retreat of the ice cap at the end of the
Pinedale Glaciation. These ice blocks became buried by **glacial outwash** and lake
sediments. The depressions left behind when the blocks melted became kettle
holes, small ponds that dot the landscape, generally along lines parallel to the
retreat of Pinedale ice. Sets of end moraines from the Pinedale Glaciation can be
seen a few miles north of the park in Montana at Eightmile Creek and near Chico
Hot Springs. Closer to the park, a deep channel that carried glacial meltwater can
be seen 5 km (3.1 miles) north of the town of Gardiner.

Since the park was covered by Pinedale glacial ice, it might seem likely that the
Grand Canyon of the Yellowstone was scoured out or that an existing canyon was
greatly deepened by the flow of ice down its length. Surprisingly, this is not the
case. Geologic evidence from in and around the canyon indicates that it was cut by
the Yellowstone River over many millennia, prior to the Pinedale Glaciation. A
glaciation that preceded the Pinedale filled the canyon with ice and protected the
canyon walls from glacial scouring, as a regional glacier flowed *across* the canyon,
rather than down it. Large glacial erratic boulders are perched on the rim of the
canyon near Artist's Point, Canyon Village (Fig. 6.7). The bedrock source for these
boulders lies to the northeast, in the Beartooth Mountains. Ice dammed the
Yellowstone River near Canyon Village, forming "Hayden Lake" in the Hayden
Valley area. Bluffs cut by Elk Inlet Creek and other streams flowing through that
valley today have exposed lake sediments formed in Hayden Lake.

Toward the end of the Pinedale Glaciation, another lake was formed, filling the
Grand Canyon of the Yellowstone. Ice flowing from the Beartooth region receded
to the Tower Falls area. The Grand Canyon of the Yellowstone was free of ice at that
time, but the lower end of the canyon was blocked by the southwest margin of
Beartooth ice. The ice dammed the Yellowstone River, forming a lake that filled the
canyon to a depth of about 180 m (590 ft). This lake has been called "Retreat Lake"

Figure 6.7. A glacial erratic boulder near Canyon Village, Yellowstone National Park. (Photograph by the author.)

because its formation was due to the retreat of regional ice. Retreat Lake nearly filled with silt before the ice finished receding and the dam melted away. The initial outlet to the lake was near Lost Lake, at an elevation of about 2100 m (6890 ft) (Fig. 6.8). When the ice receded enough so that this outlet channel was abandoned, new outlets formed at lower elevations, near Tower Junction. As progressively lower outlets were used, the canyon known as "The Narrows" was cut, and lake sediments that had nearly filled Retreat Lake were cut through by the Yellowstone River, leaving many gravel-covered terraces perched along the walls of the canyon.

As I mentioned in Chapter 5, when glacial lakes drain, the results are sometimes catastrophic. At least two large floods rushed down the drainages of the Lamar and Yellowstone rivers as Pinedale ice retreated. According to geologist Ken Pierce, the floodwaters were 45–60 m (150–200 ft) deep. About 5 km (3.1 miles) northwest of Gardiner, Montana, a flood deposit of late Pinedale age can be seen between the road and the river. It is a river channel bar deposit 20 m (65 ft) high and 450 m (1475 ft) across whose surface is covered by giant ripple marks. The downstream

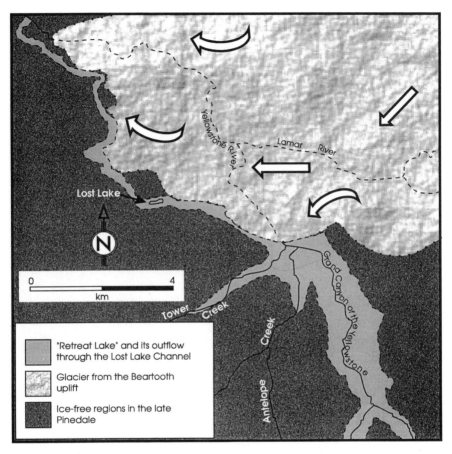

Figure 6.8. "Retreat Lake" in the lower Grand Canyon of the Yellowstone River, dammed by ice from the Beartooth uplift region. The lake filled with sediments to a depth of 180 m (590 ft). Large arrows show direction of glacial ice flow. (After Pierce, 1979.)

side of the ripple crests is littered with boulders as large as 2 m (6.5 ft) in diameter. This feature extends for about a kilometer, along with other bars and bouldery ridges. The floodwaters apparently swept up materials from moraines in the Deckard Flats region and carried them downstream. Elsewhere, late Pinedale floods have left scoured landscapes and laid down flood deposits at the mouth of Reese Creek (along the northern park boundary, west of the Yellowstone River) and at the mouth of Yankee Jim Canyon.

The ice that flowed north out of the park during the Pinedale Glaciation is called the Northern Yellowstone Outlet Glacier. This glacier exited the park at Gardiner, Montana. It left behind well-preserved scour marks and deposits on the

Figure 6.9. Glacial scour features on Dome Mountain divide, 600 m (1970 ft) above the Yellowstone River. The Northern Yellowstone Outlet Glacier filled Yankee Jim Canyon and spilled across the high pass in the upper right of the photograph, flowing toward the left side of the photograph. (Photograph by Ken Pierce, U.S. Geological Survey.)

landscape, especially where the topography and type of bedrock are favorable for the development of these features. One place where these features can be seen in the park is on Dome Mountain divide, above the Yellowstone River (Fig. 6.9). Here ice scoured the bedrock to form depressions and left mounds of debris in small ridges and hills. Exposed bedrock in this region has been polished smooth by the glacier.

Elsewhere in the park, the interaction of receding ice and hot springs created some interesting landscape features. In the vicinity of Mammoth Hot Springs, the late Pinedale glacier melted erratically, littering the landscape with large, isolated blocks of ice. These giant ice cubes remained long enough to be buried by sediments left behind by the receding ice margin. When the ice blocks finally melted, they left a series of alternating depressions and small hills, called a *kettle and kame landscape* (Fig. 6.10). Just south of Mammoth, the cone-shaped feature called Capitol Hill, and others nearby, probably formed when glacial sediments accumu-

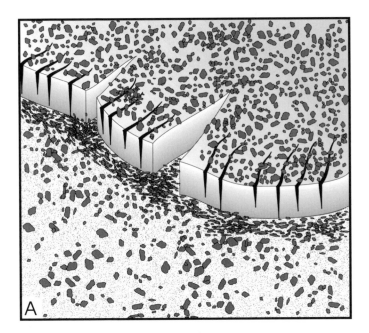

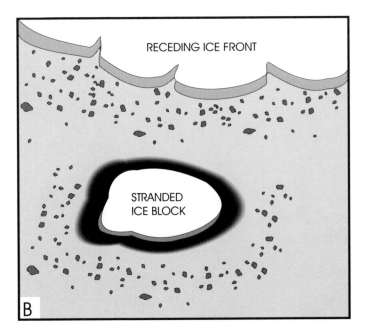

Figure 6.10. The creation of a kettle lake. (A) The Pinedale ice margin receded, leaving large volumes of glacial debris. (B) Blocks of ice were also left stranded on the landscape. (C) These blocks were buried by debris from the melting glacier. (D) When the blocks melted, they left kettle holes, alternating in some regions with mounds called kames. The kettle holes often filled with water to form steep-sided ponds.

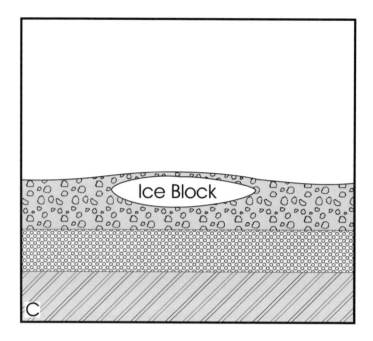

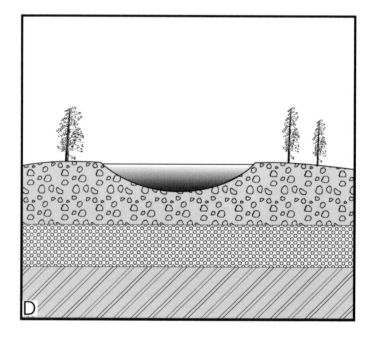

Figure 6.10. (*continued*)

lated in holes and embayments melted in the ice by the hot springs. Kettle and kame topography is common throughout the Yellowstone region, wherever blocks of ice were left stranded by receding glaciers.

The effects of the last glaciation can be seen throughout Yellowstone National Park and adjacent areas. The Yellowstone terrain visible today was shaped by a unique combination of forces: fire and ice. The fiery hot magma just beneath the surface has shaped the landscape through caldera explosions, lava flows, and geothermal features. Glacial ice scoured some regions and draped others with debris. On a geologic time scale, Yellowstone is a dynamic, ever-shifting landscape.

Late Pleistocene Vegetation History: The Forest-Tundra Two-Step

We will begin our paleoecological excursion through Yellowstone at the last, or Sangamon, interglaciation. Previous events (older than 140,000 yr B.P.) have mostly been obscured by Bull Lake and Pinedale glaciers. Geologist Richard Baker from the University of Iowa has studied pollen and plant macrofossils from lake sediments at several sites that were deposited during Sangamon and early Pinedale times. The paleoenvironmental history of the last 140,000 years is summarized in Figure 6.11. Sediments from Beaverdam Creek, near the east shore of Yellowstone Lake (Figs. 6.12 and 6.13), yielded botanical evidence of a transition from late glacial to full interglacial environments, thought to represent the onset of Sangamon climate. A cold late glacial (pre-Sangamon) climate supported tundra vegetation that gave way to forest typical of the subalpine zone today: spruce, fir, and whitebark pine. At the peak of the warm period, about 127,000 yr B.P., regional forests were dominated by Douglas-fir, with limber and ponderosa pine. These trees were growing at sites above their modern elevational limits, suggesting that the climate was warmer than it is today. As climatic cooling began again, the subalpine spruce-fir forest became dominant once more, giving way to tundra as Pinedale cooling began (Fig. 6.11)

There is a gap in pollen-bearing lake sediments from the end of the last interglacial through the early part of the Pinedale Glaciation in the Yellowstone region. The earliest indications of Pinedale environments come from the pollen record of Grassy Lake Reservoir, just south of the park boundary (Fig. 6.12). This record indicates a climatic sequence for an early Pinedale interstadial that progressed from cold to warm to cold again. The age of this interstadial is constrained by two stratigraphic horizons. It is younger than (stratigraphically above) sediments containing pollen representing full-interglacial conditions, and it is older than (stratigraphically below) the Pitchstone Plateau rhyolite lava flow, which is dated at 70,000 yr B.P. Based on these constraints, the early Pinedale interstadial

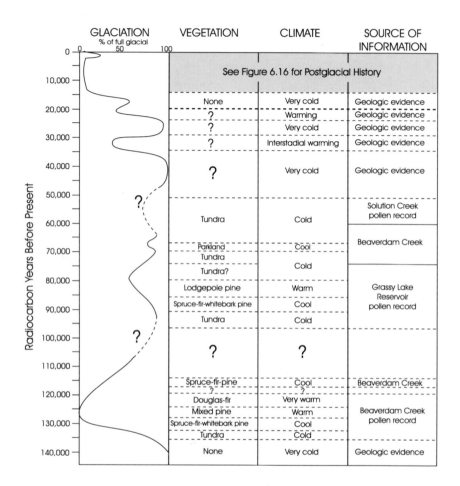

Figure 6.11. Summary of environmental history of the Yellowstone region during the last 140,000 years, based on geologic and paleontological data. (Vegetation history after Baker, 1986; glaciation history after Baker, 1986, and Sturchio et al., 1994.)

record probably began at about 95,000 yr B.P., when the vegetation sequence of tundra to spruce-fir forest was repeated. However, the warming pulse that brought about this transition was not sufficiently strong to usher in full interglacial conditions. This interstadial culminated in the establishment of lodgepole pine forest in the Yellow-stone region, from about 85,000 to 80,000 yr B.P. Lodgepole pine grows well on the type of volcanic soils found on the Central Plateau of Yellowstone, soils formed from rhyolite bedrock. During times when climatic conditions were unfavorable for the growth of lodgepole pine, the Central Plateau was essentially treeless, even though spruce, fir, and whitebark pine were growing elsewhere in the region.

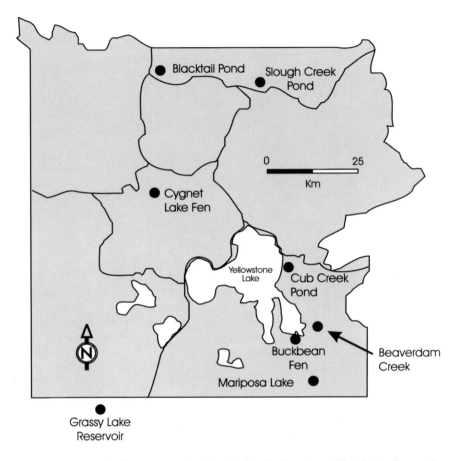

Figure 6.12. Map of Yellowstone National Park, showing locations of fossil sites discussed in the text.

Colder conditions returned about 80,000 yr B.P., and forests gave way to tundra. Another minor warming took place after 70,000 yr B.P., as indicated by increased amounts of pine, spruce, and fir pollen at the Beaverdam Creek site (Fig. 6.12). This warming lasted only a few thousand years at most, and tundra covered the landscape from about 68,000 until at least 50,000 yr B.P., based on pollen assemblages from a site at Solution Creek, studied by U.S. Geological Survey geologists Gerry Richmond and J. Platt Bradbury.

No pollen-bearing sediments have been found in the Yellowstone region from the interval from about 50,000 to 14,000 yr B.P. The park region was covered by Pinedale ice from about 30,000 to 14,000 yr B.P., but we do not know if glaciers formed between 50,000 and 30,000 yr B.P., or if the region was just cold and dry.

Figure 6.13. The southeast arm of Yellowstone Lake, viewed from the Absaroka Range, showing the region cored by geologist Richard Baker. Two Ocean Plateau Lake (foreground) is dammed by a late Pinedale moraine. (Photograph by Richard Baker, U.S. Geological Survey.)

Several late Pinedale-age sites in and around the park have provided pollen assemblages. These have been studied chiefly by Richard Baker and Cathy Whitlock. Blacktail Pond lies at just over 2000 m (6560 ft) near the northern park boundary (Figs. 6.12 and 6.14). This site was deglaciated about 14,500 yr B.P., and the pond sediments began accumulating pollen about 14,000 yr B.P. The oldest pollen assemblages indicate tundra vegetation, giving way to spruce parkland by about 12,800 yr B.P. The northern sector of the park is currently much drier than the Central Plateau and southern highlands. This has apparently held true throughout much of the late Pleistocene, if not before, as precipitation patterns are strongly tied to regional topography. So the tundra vegetation at Blacktail and Slough Creek ponds, farther east (Fig. 6.12), lacked the moisture-loving plants, such as dwarf birch and some alpine species, seen in ancient vegetation records from sites such as Buckbean Fen, farther south in the park (Fig. 6.12). In addition, late glacial forests farther south included fir and poplar, whereas the Blacktail Pond region supported only spruce parkland in the early part of the late glacial warming. The differences in climate and vegetation between north and south became more pronounced during the Holocene.

Figure 6.14. Blacktail Pond, Yellowstone National Park. (Photograph by the author.)

Starting about 12,000 yr B.P., the Yellowstone region became forested. Engelmann spruce was the first conifer to become established in most parts of the park, followed by whitebark pine and lodgepole pine. Pollen records from northwestern Wyoming suggest that Engelmann spruce migrated north into the Yellowstone region as ice retreated, at a rate of about 200 m (656 ft) per year. All of the conifers currently found in this region were apparently able to survive the last glaciation in ice-free regions of northwestern Wyoming. Unglaciated highlands south of Grand Teton National Park had favorable environmental conditions for the growth of spruce, fir, and whitebark pine.

As conditions warmed following deglaciation, the elevation of upper treeline in Yellowstone climbed 450 m (1475 ft) in 300 years, as conifer forests became established on successively higher ground. By about 9500 yr B.P. lodgepole pine had become regionally important. A combination of increasing temperature and precipitation fostered the establishment of pines from 11,800 to 9500 yr B.P. According to the Milankovitch theory, at 10,000 yr B.P. the Yellowstone region received 8.5% more solar radiation than it does today during summer and 10% less solar radiation during winter.

As I mentioned earlier, the Central Plateau region is underlain by rhyolite bedrock. The soils that develop from this parent material are relatively infertile, and lodgepole pine is the only regional tree species that does well on this type of soil. Because of these circumstances, the Central Plateau region remained treeless until about 10,000 yr B.P., when lodgepole pine invaded the region, as recorded in pollen assemblages from Cygnet Lake Fen (Fig. 6.12). After 10,500 yr B.P., the southern region of the park was clothed in forests typical of the modern subalpine regions. Increased warming by 10,000 yr B.P. allowed lodgepole pine to spread through these forests, reaching the Central Plateau, where it has been the uncontested dominant ever since.

Late Pleistocene Mammals of the Yellowstone Region

There are few known deposits containing Pleistocene vertebrate remains within the park boundaries of Yellowstone, so most of our knowledge of regional faunas comes from sites outside the park. Probably the most important of these is Natural Trap Cave, east of the park near the Montana-Wyoming border. The ancient fauna of the cave has been described by Miles Gilbert and Larry Martin of the University of Kansas. Natural Trap Cave contains sediments preserving bones ranging in age from 21,000 to 11,000 yr B.P., or characteristic of late Pinedale times. The cave fauna (Fig. 6.15) includes many extinct animals, including dire wolf, short-faced bear, American lion, American cheetah, mammoth, four kinds of extinct North American horses, American camel, woodland musk-ox, and extinct species of bighorn sheep, bison, and pine marten. The cave faunas also contain species that are no longer native to this region, including Arctic inhabitants such as the collared lemming and Arctic hare. In addition to these exotic elements, Natural Trap Cave preserves the remains of many mammals still native to northwestern Wyoming, including antelope, gray wolf, cottontail rabbit, chipmunk, pocket gopher, and several species of rodents. Most of the extinct species from these cave deposits are large mammals, whereas all of the species in the fossil assemblages that are still living today (whether in Wyoming or elsewhere) are small to medium-sized animals. There are many grazing animals in the fossil assemblages. The region east of Yellowstone was likely grassland throughout the late Pleistocene, just as it is today.

However, the variety of animal life seems to have been far richer than it is today. Imagine the fauna of the African savannah transported to a cooler climate. In Pinedale times, this image would have fit the Yellowstone region reasonably well. In place of African elephants, there were Columbian mammoths. In place of African cheetahs and lions, there were North American cheetahs and lions. Short-faced bears, gray wolves, and dire wolves hunted camels, musk-oxen, and Ameri-

Figure 6.15. A sampling of the Pinedale-age vertebrate fauna of northwestern Wyoming, based on fossil evidence from Natural Trap Cave: 1, North American lion (extinct); 2, gray wolf; 3, American cheetah (extinct); 4, antelope (pronghorn); 5, Yesterday's camel (extinct); 6, Columbian mammoth (extinct); 7, American horse (extinct); 8, caribou (no longer living in this region); 9, bald eagle (there were no eagle remains in the cave, but its presence in the region seems likely).

can horses, as well as antelope and bison. By 11,000 yr B.P., many of these animals no longer existed. Their Old World relatives have managed to survive through the Holocene, but here they became extinct.

Yet the comparisons with the modern fauna of Africa may be misleading, because the Pinedale climates of northern Wyoming were very different from those of the African savannah. There is a strong arctic-subarctic element in the

Pinedale faunas of Wyoming. In addition to collared lemming and arctic hare, musk-oxen and caribou also ranged across the grasslands of Wyoming during the late Pleistocene. These species are found only in Alaska and northern Canada today. So while the plains of Wyoming may have been as dry as the modern African savannah, they were certainly far colder.

The extinction of large mammals in Wyoming at the end of the Pleistocene was part of a continental-scale disaster. All the proboscidians (mammoths and mastodons), the horses, camels, giant sloths, and many other species became extinct in North America at the end of the last glaciation. Why did this happen? The obvious answer might seem to be that these cold-adapted animals could not tolerate the warm climates of the Holocene. This argument might be convincing if it weren't for the fact that the same cold-adapted species and their ancestors had withstood the warm climates of about a dozen previous interglacial periods, at least one of which was probably substantially warmer than anything yet experienced in the Holocene!

No, climatic warming per se was not enough to extinguish the **megafaunal mammals.** Some unique biological or environmental factors must have influenced megafaunal populations at the beginning of the Holocene. Some argue that human beings were the most important agent in dispatching the North American megafauna. Paul Martin, geologist at the University of Arizona, coined the phrase *Pleistocene overkill* to describe how people may have hunted these animals to extinction. The theory suggests that the megafauna of North America was especially vulnerable to late Pleistocene (Paleoindian) hunters because the people were newcomers on this continent at that time, and the animals, unaccustomed to human hunters, had little natural fear of them. The theory holds that this new hunting pressure, combined with rapid climate change, wiped out most of the megafauna on this continent. We will probably never know if the overkill theory is the right one, since the fossil evidence is so spotty. Whatever the cause of the extinction, we are left with a rather poor collection of large animals that have made it through the Holocene. Nevertheless, they are true survivors. If they were human, they would probably sport hats and T-shirts with the motto, "We Survived the Pleistocene-Holocene Transition"!

Holocene Ecosystems

During the last 10,000 years (the interval termed the Holocene by geologists), changing climates in the Yellowstone region brought about some large-scale changes in regional vegetation (Fig. 6.16). Today the northern part of the park is considerably drier and warmer than the highlands to the south. This is easily appreciated in late

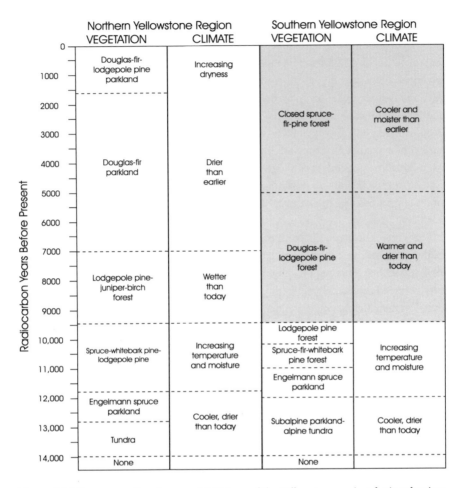

Figure 6.16. Summary of environmental history of the Yellowstone region during the time since the last (Pinedale) deglaciation. Shaded section indicates the time interval when the environment of the southern park region was substantially different from that of the northern region. (After Whitlock, 1993.)

spring, when the southern parts of the park are buried under meters of snow while the Mammoth region is often snow free—the "banana belt" of Yellowstone.

Thanks in large part to the work of palynologist Cathy Whitlock, we have come to understand how topographic differences between these regions affected Holocene environments. During the early Holocene (9500–7000 yr B.P.), the northern part of the park experienced wetter climates than it does today, whereas the southern region was warmer and drier than it is today. Paleoclimate researchers believe that these changes were brought about by shifts in atmospheric circulation, linked with shifts in precipitation patterns. Paleoclimate reconstructions suggest that the increased summer insolation during the early Holocene created summer weather patterns dominated by high pressure in the southern Yellowstone region. Relatively warm and dry conditions tend to increase fire frequency in this region, and increased numbers of forest fires helped to maintain such fire-adapted species as lodgepole pine, Douglas-fir, and aspen. These trees outcompete other species in fire-prone regions. For instance, the cones of lodgepole pine are triggered to release their seeds when they are heated by forest fires. This characteristic ensures that a large crop of lodgepole pine seedlings will sprout in recently burned landscapes, overwhelming the seedlings of other trees. Thus the southern Yellowstone region supported forests dominated by lodgepole pine and Douglas-fir from 9500 to 5000 yr B.P.

The same atmospheric conditions that fostered a warm and dry climate in southern Yellowstone during the early Holocene brought increased moisture to the northern sector. This region, plus parts of central and eastern Wyoming, apparently received increased precipitation from summer monsoons. Today the monsoonal system brings Pacific moisture to the southwestern United States. From 9500 to 7000 yr B.P., the northern Yellowstone region supported forests of lodgepole pine, juniper, and birch. The region probably looked quite different than it does today; we see broad parklands of grasses and sagebrush with conifers growing mostly on moister hillsides.

This pattern broke down about 7000 years ago, as the northern region became drier and the vegetation opened up into Douglas-fir parkland. This increasing aridity culminated in the modern Douglas-fir–lodgepole pine parkland ecosystem by about 1600 yr B.P. By 5000 yr B.P., the southern region began receiving increased moisture, and the modern closed spruce-fir-pine forest became established.

Early Peoples of Yellowstone

The earliest human visitors to the Yellowstone region who left any evidence of their stay are known as Paleoindians. The Paleoindian period lasted from about 11,500 to 8000 yr B.P. (Table 6.1). Paleoindian hunters and their artifacts are discussed in a book by George Frison and a book edited by Dennis Stanford and Jane Day. Archaeologists studying the way of life of these people believe that they were hunter-gatherers who moved camp fifty to a hundred times a year in search of game animals. Although they probably used campsites repeatedly through the years, they left little evidence of their presence.

Based on information gathered from their descendants in the late nineteenth century, anthropologists believe that the Paleoindians of the Yellowstone region

Table 6.1. Northern Plains Indian Cultural Chronology

Culture	Years before Present
Paleoindian	11,500–8000
Early Plains Archaic	8000–5000
Middle Plains Archaic	5000–2700
Late Plains Archaic	2700–1500
Late Prehistoric	1500–European contact (about 200 yr B.P.)

followed certain seasonal patterns of activity. In early spring they lived on the plains to the east and south of the Yellowstone uplands, gathering tender shoots and greens. However, based on the sources of stone used in tools by these people, they probably also had extensive contact with regions west of the park, where root and bulb gathering was an important early spring activity.

In late spring, berries, roots, and bulbs were gathered. In summer, the Paleo-indians entered the mountain regions, where they gathered berries and pine cones. Seeds (pine nuts) were removed from the cones and stored as food for winter. In winter and spring, they hunted buffalo and other large game in the lowland regions. In summer, they hunted mainly bighorn sheep, although mule deer were also a major food source for the peoples in the foothills and mountain regions.

The hunting of bighorn sheep deserves special attention here, because the archaeological evidence suggests that this activity was an essential part of the lives of the ancient peoples of the Yellowstone region. There are numerous game drive walls in the Absaroka Mountains, where hunting parties funneled groups of sheep to an ambush point. The remains of nets have been found at some of these sites. Bighorn sheep become docile after being netted, whereas other large game animals struggle fiercely to get away. Once the animals were trapped in the net, hunters could walk up and club them. This method of bighorn sheep hunting continued from Paleoindian through Late Prehistoric and Early Historic times (Table 6.1). Bighorn sheep remains are the dominant fossil group in many regional cave deposits, including those in Mummy Cave, near Cody, Wyoming. Over 13,000 bones and bone fragments have been recovered from this cave; the majority of these are from bighorn sheep.

By 10,000 years ago, the people of the Yellowstone region had become separated into two cultural groups: foothills-mountain people and plains-basins people. The highlands group hunted bighorn sheep, mule deer, and, to a lesser extent, bison. The lowlands group developed communal bison hunting techniques, using both natural landforms (such as cliffs and ravines) and corrals to capture animals. The two groups developed distinctive styles of **projectile points.** It appears that the members of the highland group were relatively isolated, for there was more local development of styles of points in this region. It was possible for a given group of hunters to exploit either region (highlands or lowlands), but not both. The foot-hills-mountain people enjoyed more diverse food resources than the plains-basins people. Besides bighorn sheep, mule deer, and bison, archaeologists have also found the remains of many smaller mammals and diverse assemblages of plant foods in sites occupied by these people.

Some archaeologists believe that drought conditions about 8000 yr B.P. forced the lowlands group to retreat back up to the mountains. Whether or not this is so, regional peoples had by this time developed new styles of projectile points; on the plains of Wyoming, this culture is termed Early Plains Archaic (Table 6.1). In the Yellowstone region, the term Early Archaic may be more appropriate.

The chronology of archaeological sites within Yellowstone National Park begins with Paleoindian occupation, shortly after 12,000 yr B.P. There are several Paleo-indian sites in the region, including Mammoth Meadows (in southwestern Mon-tana), Mummy Cave, and Indian Creek in west-central Montana. The shores of Boulder Lake, near Pinedale, Wyoming, yielded Paleoindian artifacts made from obsidian that came from Obsidian Cliffs in Yellowstone Park (Fig. 6.17).

The obsidian story, worked out by archaeologist Ken Cannon and colleagues, reveals some fascinating details of the lives of ancient Yellowstone peoples. Cannon has been tracking down the obsidian that became worked into points, blades, and other tools. Although obsidian glass may be more easily broken than other rock types used for toolmaking (such as chert and flint), there is nothing sharper than the edge of an obsidian blade. These blades are considerably sharper than the most finely honed surgical steel; indeed some surgeons prefer obsidian scalpels over steel scalpels for this very reason. Obsidian tools lend themselves to archaeological study for two reasons. First, the source of the obsidian can be readily determined because each obsidian deposit is chemically unique; thus the source locality of artifacts can be found through chemical fingerprinting. Second, through the chemical dating method known as the obsidian hydration technique, the age of obsidian artifacts can be determined, even if they are not buried in stratigraphic layers that contain charcoal or other organic materials suitable for radiocarbon dating.

Cannon and his research team from the National Park Service have collected thousands of obsidian artifacts in and around the park (Fig. 6.18). They have tracked down the sources of 500 of these artifacts. The oldest tool known to have been made from Obsidian Cliff glass is a Folsom point, found on federal land near the Bridger-Teton National Forest. The point dates to between 10,900 and 10,200 yr B.P. Obsidian was apparently highly valued for toolmaking. Extensive trade

Figure 6.17. Obsidian Cliffs, Yellowstone National Park. Small bands of obsidian glass occur throughout the volcanic rocks at this site. (Photograph by the author.)

networks developed over time, and in Late Prehistoric times obsidian from Obsidian Cliffs found its way as far east as Ohio, where obsidian artifacts were placed in burial mounds of the Mound Builder (Hopewell) culture. Some Paleoindian artifacts found in the park have been traced to obsidian outcrops from sites at Bear Gulch, Idaho, and Teton Pass, Wyoming. Bear Gulch is about 360 km (225 miles)

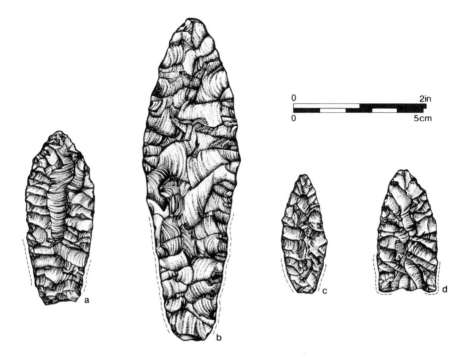

Figure 6.18. Paleoindian projectile points from Yellowstone National Park, made of obsidian. (A) An Agate Basin point, sourced to Obsidian Cliff. (B) A Hell Gap point, sourced to Bear Gulch, Idaho. (C) A Late Paleoindian lanceolate point, sourced to Teton Pass, Wyoming. (D) A reworked late-Paleoindian lanceolate point, sourced to Bear Gulch, Idaho. (Illustration by Janet Robertson, Midwest Archeological Center, National Park Service.)

southwest of Yellowstone, and Teton Pass is about 175 km (110 miles) south of the park. As might be expected, early- and mid-Holocene-age obsidian artifacts in the park come mostly from Obsidian Cliffs, whereas late Holocene artifacts show more diverse obsidian sources, reflecting the increasing development of regional trade.

Artifacts made of obsidian from outside the park provide information on human migration patterns and seasonal movements. There appears to have been a well-traveled seasonal circuit that included stops in the Jackson Hole region, the Snake River Plain, the Centennial Mountains, and Yellowstone. Many of the obsidian artifacts found in the Yellowstone region were probably discarded or lost by traveling bands of Paleoindians. They are mostly broken or used up (i.e., they are pieces of tools that have been sharpened and resharpened until they are no longer usable).

Based on the numbers of sites and artifacts, regional human populations increased markedly in the late Holocene, with substantial occupations after 4500 yr B.P. (Middle Plains Archaic times). Further increases in population took place in the last 2000 years. The folklore of the American West would have it that the Indian tribes of northwestern Wyoming scarcely ever visited the Yellowstone region because they were afraid of the geysers, fumaroles, and other geothermal features. Archaeological research during the last few years has proven otherwise. The regions we call wilderness have been called home by indigenous peoples since the end of the last ice age.

Yellowstone is, by its very nature, a land given to tall tales. When trapper Jim Bridger visited the region early in the nineteenth century, he came back with only slightly exaggerated stories of petrified trees, geysers, and the like. Given the typical mountain man's habit of trying to top his cronies in storytelling, these tales were generally written off by his friends and business associates. It was only when other, more reputable reporters made the trip to Yellowstone that Bridger was vindicated.

Suggested Reading

Baker, R. G. 1986. Sangamonian(?) and Wisconsinan paleoenvironments of Yellowstone National Park. *Geological Society of American Bulletin* 97:717–736.

Cannon, K. P. 1993. Paleoindian use of obsidian in the greater Yellowstone area. *Yellowstone Science,* Summer 1993: 6–9.

Frison, G. C. 1991. *Prehistoric Hunters of the High Plains.* 2d ed. San Diego, California: Academic Press. 532 pp.

Fritz, W. J. 1985. *Roadside Geology of the Yellowstone Country.* Missoula, Montana: Mountain Press. 149 pp.

Gilbert, B. M., and Martin, L. D. 1984. Late Pleistocene fossils of Natural Trap Cave, Wyoming, and the climatic model of extinction. In Martin, P. S., and Klein, R. G. (eds.), *Quaternary Extinctions: A Prehistoric Revolution.* Tucson: University of Arizona Press, pp. 138–147.

Jackson, L. E., Jr., and Clague, J. J. 1991. The Cordilleran Ice Sheet: One hundred and fifty years of exploration and discovery. *Géographie physique et Quaternaire* 45:269–280.

Pierce, K. L. 1979. History and Dynamics of Glaciation in the Northern Yellowstone National Park Area. U.S. Geological Survey Professional Paper 729-F. 90 pp.

Stanford, D. J., and Day, J. S. (eds.). 1992. *Ice Age Hunters of the Rockies.* Niwot: University Press of Colorado. 378 pp.

Sturchio, N. C., Pierce, K. L., Murrell, M. T., and Sorey, M. L. 1994. Uranium-series ages of travertines and timing of the last glaciation in the northern Yellowstone area, Wyoming-Montana. *Quaternary Research* 41:265–277.

Whitlock, C. 1993. Postglacial vegetation and climate of Grand Teton and southern Yellowstone National Parks. *Ecological Monographs* 63:173–198.

Whitlock, C., and Bartlein, P. J. 1993. Spatial variations in Holocene climatic change in the Yellowstone region. *Quaternary Research* 39:231–238.

7

GRAND TETON NATIONAL PARK
A Landscape Sculpted by Ice

One of the most startling sights to be seen in North America comes into view as you drive north on the highway from Jackson toward Grand Teton National Park. You round a single bend at the end of this gentle climb—and there are the Tetons, towering over you as they stretch more than 2000 m (6560 ft) into the sky! The Teton Range is relatively small, only a few tens of miles from north to south, but this range presents probably the most dramatic mountain front in the region, because the tall mountains rise directly from the plains below, with no intervening foothills (Fig. 7.1). This is because the Tetons are the product of **block faulting,** a process in which huge blocks of bedrock move in opposite directions along **fault** lines. In this case, the Tetons form a block that is being uplifted along the west side of the fault, while the Jackson Hole region represents an eastern block that is dropping in elevation. The difference in elevation between the two blocks is thought to be as much as 9150 m (30,000 ft). The surface of the Jackson Hole block is mantled by thick layers of sediments that have accumulated over the last few million years. Most of the uplift of the Tetons has taken place in perhaps the last five million years, so the Tetons are very young mountains, geologically speaking. "Young" mountains are usually more spectacular to our eyes than "old" mountains, because young mountains have steep slopes, sharp ridges, and pointed peaks.

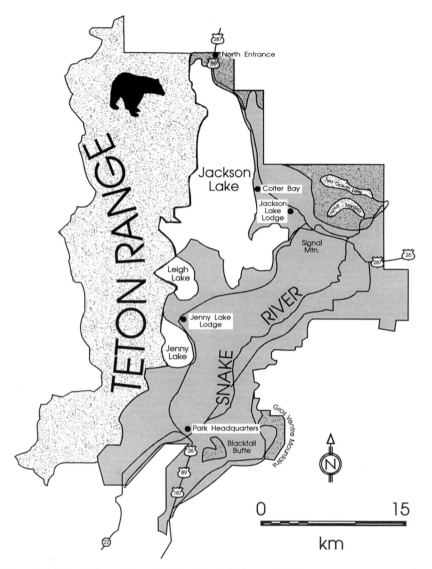

Figure 7.1. Map of Grand Teton National Park. Light, speckled areas represent mountains; gray areas represent lowlands.

Old mountains, such as the Appalachians, are worn down; their peaks are more mound shaped, their ridges more gentle, and their slopes more gradual.

The process of mountain building is still going on in the Tetons. Movements along the fault measure as much as 34 m (110 ft) during the last 14,000 years. These shifts are likely to continue far into the future.

Modern Setting

The modern ecosystems of Grand Teton National Park are similar in most ways to those of Yellowstone (Fig. 7.2). Sagebrush steppe dominates the lowlands of Jackson Hole, up to an elevation of about 1500 m (4900 ft). Pine parkland ranges from 1500 to 2400 m (7875 ft) in elevation. In the lower elevations of this range, limber pine is the dominant tree. This gives way to Douglas-fir in a narrow elevational band between about 1900 and 2000 m (6230–6560 ft), then lodgepole pine dominates up to the boundary with subalpine forest at 2400 m (7875 ft). Subalpine forests in the park include Engelmann spruce, subalpine fir, and whitebark pine. Whitebark pine forms the upper treeline at about 2850 m (9350 ft). Above this elevation is an alpine tundra zone, with low-growing shrubs, herbs, and sedges.

The Snake River dominates the plains in the shadow of the Tetons, called Jackson's Hole after trapper David Jackson (the early fur trappers referred to broad, flat valleys as "holes"). Wildlife flourishes along the course of the Snake. Osprey and bald eagles nest along its banks in dead trees. Otters hunt fish and frogs, and the small ponds created by beavers offer the succulent aquatic vegetation favored by moose. The sagebrush steppe is grazed by bison and pronghorn antelope, while the lodgepole pine forests above the plain are home to elk, mule deer, and black bear. Elusive animals, such as the mountain lion, lynx, pine marten, and bobcat, find refuge from human disturbance in the rugged Teton wilderness.

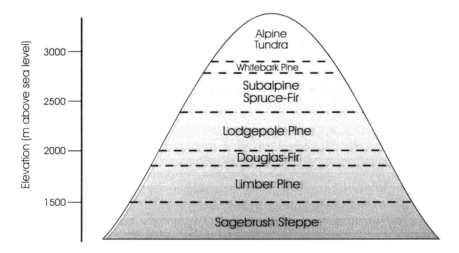

Figure 7.2. Modern ecosystems of Grand Teton National Park. (After Whitlock, 1993.)

Glacial History

Superimposed on the block-faulted features of the Tetons are the effects of multiple Pleistocene glaciations. Glacial ice flowed out of high mountain cirques to carve valleys, trim mountain walls, and deposit moraines and other features on the valley floor below. The following reconstruction of glacial events is based on a recent report for the U.S. Geological Survey by Ken Pierce and John Good. The Bull Lake Glaciation, about 140,000 yr B.P., was the most extensive of the late Pleistocene. In the Jackson Hole region, this glaciation is called the Munger Glaciation, after Munger Mountain at the south end of Jackson Hole. A tongue of ice about 1000 m (3300 ft) thick flowed south from the Yellowstone region, along the eastern mountain front of the Tetons. This ice buried the Gros Ventre buttes, Blacktail Butte, and Signal Mountain. Glaciers flowed down the Gros Ventre valley from the Wind River Range, between the Gros Ventre Mountains and the Teton Range, and then into the Snake River Canyon. The Munger Glaciation filled Jackson Hole with ice, and Munger moraines more than 100 m (330 ft) thick are located in the southern part of Jackson Hole. The glaciers that flowed south out of Yellowstone were much more substantial than the mountain glaciers that formed in the Tetons. This is because the Tetons are a much smaller source area for the accumulation of snow and ice, especially in comparison to the highlands to the north, where large regions lie above 2700 m (8860 ft). Large glaciers also came down the Pacific Creek and Buffalo Fork drainages. During Pleistocene glaciations, more snow accumulated in winter than melted in summer at elevations of 2700 m or greater in the northern Rockies.

During the Little Ice Age (about A.D. 1500 to 1900), mountain glaciers advanced in many regions of the Rockies, including the Glacier National Park region. But historical photographs, such as the one of the Grand Teton taken by W. H. Jackson in 1872 (Fig. 7.3), show that the Teton Range experienced hardly any buildup of ice. Apparently, the same factors that kept Teton glaciers small in Pinedale times also kept them small during the Little Ice Age.

After years of intensive study, Pierce and Good developed a history of glacial events during Pinedale times, i.e., the interval from 80,000 to 10,000 yr B.P. The patterns of ice ebb and flow are complicated, as is the sorting out of ice sources and the age of glacial moraines (end points). One complicating factor is the fault activity along the Teton–Jackson Hole front. Deposits of Pinedale outwash (sediments that washed out from the margins of retreating ice fronts), aided by down-faulting on the Teton fault, have built up deposits more than 60 m (200 ft) thick in the Pinedale end moraine–outwash regions, burying the landscapes that existed during Pinedale ice advances. So, in order to understand what happened during Pinedale times, it is necessary to look at surfaces

Figure 7.3. Photograph of the Grand Teton taken by W. H. Jackson in 1872. Note the lack of substantial glaciers on the mountain. (Photograph courtesy U.S. Geological Survey.)

that have since been buried under thick mantles of sediment. Needless to say, this is not an easy thing to do.

During the Pinedale interval, lobes of ice from three separate sources flowed into the Teton–Jackson Hole region, in addition to mountain glaciers from the Tetons themselves. As in the Munger Glaciation, these mountain glaciers were far smaller than the glaciers coming from other regions. Valley glaciers originating in the Tetons extended only about 2.5 km (1.5 miles) out beyond the mountain front, forming end moraines that dammed Jenny, Bradley, Taggart, and Phelps lakes. North of Jenny Lake, glaciers from the Tetons joined ice flowing south in the Snake River lobe. The Snake River lobe entered this region from the north, while the Buffalo Fork lobe flowed into Jackson Hole from the east, and the Pacific Creek lobe came from the northeast. The reconstruction of glacial events in the Pinedale is complicated by the fact that the size of these three lobes kept changing.

The oldest phase of the Pinedale Glaciation in Grand Teton National Park is called the Burned Ridge phase (Fig. 7.4). Although this phase has not been precisely dated, it may have occurred about 65,000 yr B.P. During this phase, ice from the Buffalo Fork and Pacific Creek lobes joined and flowed westward into Jackson Hole. The Buffalo Fork lobe was the largest lobe of the three during this phase. Three separate ice advances of this phase left moraine and outwash deposits, although the outwash has buried the end moraines either partially or totally. Major outwash deposits of the Burned Ridge phase are found in the Antelope Flats region of the park.

Ice in the Buffalo Fork and Pacific Creek lobes excavated deep depressions in the valley, called scour basins. The Buffalo Fork lobe dug a scour basin more than 20 km (12.4 miles) long, extending from Deadmans Bar up the Snake River to Moran Junction. From there, this scour basin ran up the Buffalo Fork drainage at least to Rosies Ridge. At its widest point, this depression was about 6 km (3.7 miles) wide, and when the Burned Ridge phase was over this depression filled with water to form a lake. Another scour basin at least 16 km (10 miles) long extends down Pacific Creek, west along the north side of Signal Mountain to the Spalding Bay region. This depression filled with almost 200 m (660 ft) of unconsolidated sediments beneath the northern dike of Jackson Lake dam. These sediments were determined to be too unstable to support the dike during an earthquake, and the dam had to be rebuilt in the 1980s because of this. The basins that contain Emma Matilda and Two Ocean lakes were also excavated by the advancing ice of the Pacific Creek lobe during the Burned Ridge phase.

The second of the three glacial phases in Grand Teton National Park during Pinedale times is called the Hedrick Pond phase (Fig. 7.5). This phase probably began around 30,000 yr B.P. The Pacific Creek lobe was the dominant lobe of this phase, and the Buffalo Fork lobe was considerably smaller than it had been in the Burned Ridge phase. During the Hedrick Pond phase, the Snake River lobe advanced south of the Jackson Lake region, where it joined the Pacific Creek lobe, and flowed into the lake basin scoured out during the Burned Ridge phase. This glacial phase left kettle holes on the east and west sides of Burned Ridge. It also left a large outwash fan, emanating from the head of the Spalding Bay–West Channelway.

Some glacial features in the park were formed during either the Hedrick Pond phase or the last phase of the Pinedale Glaciation, called the Jackson Lake phase.

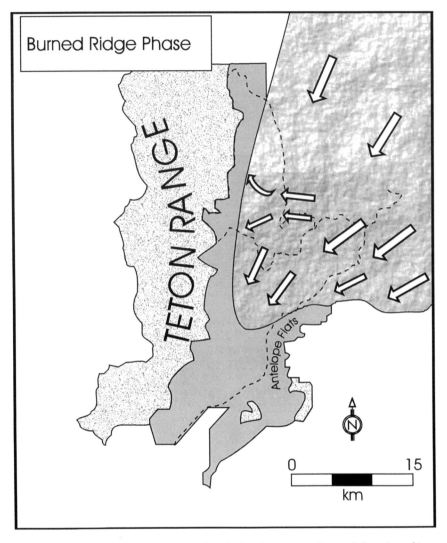

Figure 7.4. Map of Grand Teton National Park, showing extent of ice and direction of ice flow during the Burned Ridge phase of the Pinedale Glaciation. (Data from Pierce and Good, 1992.)

The Snake River ice lobe flowed south and excavated the Jackson Lake basin in Hedrick Pond and Jackson Lake times. The material that was dug out of the basin by advancing ice was deposited farther south; some of it ended up in the outwash fan at Spalding Bay–West Channelway, and some of it ended up at the region known as Potholes Channelway. Glaciers coming out of the Tetons from Leigh,

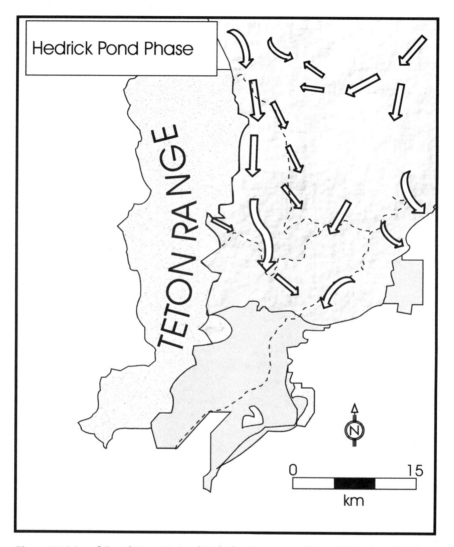

Figure 7.5. Map of Grand Teton National Park, showing extent of ice and direction of ice flow during the Hedrick Pond phase of the Pinedale Glaciation. (Data from Pierce and Good, 1992.)

Moran, and Berry canyons joined the Snake River ice lobe and flowed along its western flank to deposit terminal moraines between String Lake and the Spalding Bay–West Channelway.

There is more clearly defined evidence for some events that took place during the Jackson Lake phase (Fig. 7.6), which began before 20,000 yr B.P. and ended by about 15,000 yr B.P. This phase is named for moraines that flank the south side of

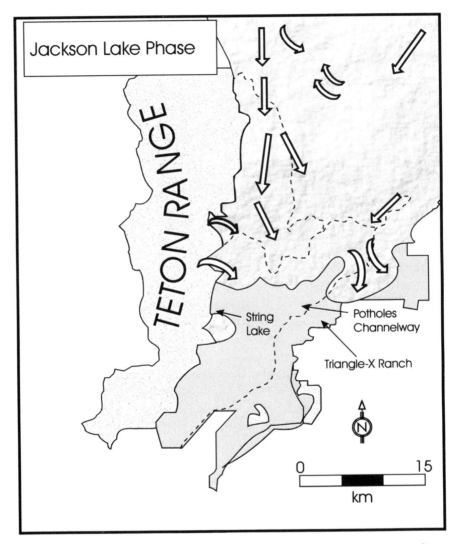

Figure 7.6. Map of Grand Teton National Park, showing extent of ice and direction of ice flow during the Jackson Lake phase of the Pinedale Glaciation. (Data from Pierce and Good, 1992.)

the lake. The terminal moraines of the Jackson Lake phase are generally well inside those of the Hedrick Pond phase, so this last glacial pulse appears to have been less extensive than the earlier Pinedale ice advances.

When Hedrick Pond ice retreated, it left behind the basin previously excavated in Burned Ridge time at Triangle-X Ranch; this basin formed a lake. In this region, outwash deposits from the Jackson Lake phase formed three large fans that **prograded** into the lake. The evidence for this can be seen today along bluffs of the Snake River, where lake sediments are overlain by gravels from these fans. The outlet of the lake was near the Snake River overlook. The outlet was apparently eroded by floodwaters from a lake dammed by the Pacific Creek ice lobe. This lake occupied the Buffalo Fork Valley during the Jackson Lake phase and near the end of the Hedrick Pond phase. As we have already seen, lakes dammed by ice tend to drain catastrophically. This lake's ice dam failed repeatedly, as the water level of the impounded lake reached 90% of the thickness of the glacial ice. Flood debris from these drainage events covers large regions downstream from the Snake River overlook, as far south as Jackson Hole airport. These floods were enhanced in a domino effect, as floodwaters from the Buffalo Fork Valley eroded the outlet of the Triangle-X lake, so that the waters of two lakes flooded down the valley.

The Snake River ice lobe retreated rapidly from its maximum extent near the southern shore of Jackson Lake. The ice margin did not remain in place long enough to deposit well-defined moraines along its retreating margin. The reason for the rapid retreat of this ice lobe is that one of the main sources of ice feeding this lobe, the Yellowstone Ice Cap, was itself stagnating and melting at a rapid rate. According to the work of Pierce and Good, by 15,000 yr B.P., the Jackson Lake phase was finished, and the Jackson Lake region was free of ice.

A cursory glance at a map of Grand Teton National Park shows that the Snake River was in the way of these various advancing and retreating ice lobes in Pinedale times. Geologic evidence indicates that the course of the river did indeed change several times, in response to changing glaciers. Between the Burned Ridge and Hedrick Pond phases, it appears that the Snake River flowed along the Teton mountain front. The present river course is the product of Pinedale glacial history, as the water follows topographic depressions left by the series of ice advances and retreats. All of Jackson Hole is mantled in outwash debris from the three Pinedale phases. This glacial fill reaches much greater heights above the modern drainage in the north, tapering southward. At Burned Ridge, it is about 90 m (300 ft) above the Snake River, whereas west of the town of Jackson, it is only 3 m (10 ft) above the river. Glacial outwash more than 60 m (200 ft) thick dams the southern margin of Jackson Lake, blocking a potential stream outlet there. So the Snake River flows out of the lake along its eastern shore, through a scour basin excavated by a westward-flowing Pacific Creek ice lobe. Just below the junction of the Snake with Pacific Creek, the river crosses a bedrock threshold. This threshold has been eroded to a low, narrow gap by scouring of ice from the three ice lobes during the three phases of the Pinedale Glaciation. From there, the river enters the Triangle-X lake basin. Its exit from this basin follows a seam between two Pinedale outwash fans. The fan on the west side is from the Hedrick Pond phase, and the fan on the east side is from the Burned Ridge phase.

On the east side of Jackson Hole, south of the Yellowstone-Absaroka Pinedale ice margin, the Gros Ventre River, Ditch Creek, and Flat Creek deposited large alluvial fans of sediments churned up by Pinedale ice. These fans fill the lowlands around Blacktail Butte, East Gros Ventre Butte, and West Gros Ventre Butte.

Glacial Features Easily Viewed in the Park

Nearly every landscape that can be seen in Grand Teton National Park has been affected by Pinedale and Munger glaciers. The advancing and retreating ice lobes acted like giant steam shovels, gouging out depressions in one region and depositing the debris from these excavations elsewhere. Many of these glacial features trend north-south, following the direction of glacial advances and retreats. The exceptions are features such as the terminal moraines damming Jenny Lake (Fig. 7.7) and Taggart Lake. These were produced by mountain glaciers and are relatively small features compared with Jackson Lake and other features shaped by the larger ice lobes that originated in the Yellowstone-Absaroka region. In order for mountain glaciers to become really large, the mountain range must be relatively large

Figure 7.7. Jenny Lake, showing lateral and terminal moraines that dammed the lake near the end of the Pinedale Glaciation. (Photograph by the author.)

(i.e., it must be a large catchment region for moisture), and the region must receive ample snowfall. The Tetons are a relatively small range, and apparently during the late Pleistocene this region did not receive sufficient snowfall to build large glaciers. The climate of the Teton region may have been greatly affected by the Yellowstone Ice Cap during Munger and Pinedale times.

If you arrive at Jackson by airplane, you land on glacial outwash of the Jackson Lake phase (the airport runways are built on this surface). Driving north through the park, you will come to the Glacier View turnout. At this locality, 8 km (5 miles) south of the limit of Pinedale ice, you may view Pinedale moraines to the west, along the flanks of the Tetons (Fig. 7.8). These moraines dammed Bradley and Taggart lakes. At the Teton Point turnout, you are standing on outwash of Hedrick Pond age, from which outwash of Burned Ridge age can be seen to the northeast. Looking along the length of the Snake River, the terrace of Jackson Lake age converges with the terrace of Hedrick Pond sediments on which you are standing, then the Hedrick Pond–age sediments rise above the Jackson Lake terrace near Moose (Fig. 7.9). If you continue looking up the Snake River drainage, you will see moraines of Burned Ridge age.

At the Snake River overlook you are standing on outwash from the third ice advance of the Burned Ridge phase. Moraines of this age rise above the outwash in this vicinity, across the highway and across the river (Fig. 7.10). The ice margin of

Figure 7.8. Moraines viewed from the Glacier View turnout, Grand Teton National Park. (Photograph by the author.)

Figure 7.9. Glacial outwash terraces seen by looking north from the Teton Point turnout, Grand Teton National Park. (Photograph by the author.)

Figure 7.10. View of moraines and outwash terraces looking southwest from the Snake River overlook, Grand Teton National Park. (Photograph by the author.)

Figure 7.11. The Jackson Lake moraine, visible 200 m (660 ft) northwest of the Cathedral Group turnout, Grand Teton National Park. (Photograph by the author.)

the second advance during the Burned Ridge phase lies 2.6 km (1.6 miles) south, and the margin of the third advance is 3.5 km (2.2 miles) south.

At the Cathedral Group turnout, moraines of the Jackson Lake phase can be seen about 200 m (660 ft) north from the turnout. The moraines are covered by forest but are large, clearly visible mounds (Fig. 7.11). At the nearby String Lake parking area, hike the trail across a bridge over String Lake Creek. Upstream at this point is a line of boulders marking an outer moraine of Jenny Lake. As you continue on the trail, this moraine forms the ridge nearest String Lake, and the trail crosses the inner moraines around Jenny Lake. A hike of an additional 500 m (0.3 mile) will bring you to a small kettle hole, a depression left by the melting of a debris-covered ice block.

At the Potholes turnout, hike south of the road (away from Spalding Bay, Jackson Lake) and you will encounter kettle and kame topography formed along an ice margin of the Pacific Creek lobe during the Hedrick Pond phase.

There are several interesting glaciological features in the neighborhood of Jackson Lake Lodge. To the southeast (between the lodge and Highway 287), you will encounter low mounds and ridges that are kame gravel deposits left by the retreating margin of an ice lobe at the end of the Jackson Lake phase. East of the lodge, between Christian Lake and Emma Matilda Lake, is a pitted kame surface that was deposited between the Snake River and Pacific Creek lobes (Fig. 7.12).

Figure 7.12. Kame gravel deposits (now vegetated) by the Emma Matilda Lake trail, south-east of Jackson Lake Lodge, Grand Teton National Park. (Photograph by the author.)

Immediately north-northwest of the lodge complex is a moraine left by the receding Snake River lobe at the end of the Jackson Lake phase (Fig. 7.13).

After the Ice Melted: The Return of Life

Palynologist Cathy Whitlock has studied pollen from lake sediments in Hedrick Pond, as well as lakes in the Teton Wilderness region northeast of Grand Teton National Park (Fallback Lake, Lily Lake and Fen, Divide Lake, and Emerald Lake) and nearby lakes in the southern Yellowstone region (Fig. 7.14). She has made the following reconstructions of regional vegetation. Prior to 11,500 yr B.P., unglaciated regions of Grand Teton National Park supported the growth of vegetation resembling but not quite matching modern alpine tundra. This vegetation included willow, birch, and juniper shrubs, abundant sagebrush, and grasses. There was more sagebrush in the vegetation than is seen today in the alpine zone. This feature suggests that the late Pinedale vegetation of the region was a mixture of cold-steppe grassland and alpine tundra elements, unlike any seen in this region today but not unlike that in some regions of Siberia and northern Alaska, where steppe-tundra vegetation dominated large regions during the Pleistocene (see Elias, 1995). The sage pollen may have come from woody sagebrush species (the

Figure 7.13. Recessional moraine formed during the Jackson Lake phase of the last glaciation by the Snake River lobe, northwest of Jackson Lake Lodge, Grand Teton National Park. (Photograph by the author.)

ones that dominate the lowlands of the park today), or it may have come from the alpine sage, which is a herbaceous (nonwoody) plant that grows only at high elevations.

The juniper shrubs growing in this region in late Pinedale times became established on the well-drained soils of outwash plains and near stagnant ice lobes. This plant was all the more successful because of the lack of competition from other conifers. The birch and willow shrubs identified in the late Pinedale pollen assemblages probably grew along rivers, streams, ponds, and lakes.

From a landscape perspective, deglaciation is a messy process. The receding ice front leaves great piles of rubble, and millions of tons of finer-grained sediments come pouring out from underneath and alongside the melting ice, as torrents of sediment-choked water are liberated from the rapidly melting ice. Even though stream flow did not always increase much beyond what we see today, all of the water that had been locked up in the ice had to go somewhere. It had to make its way down drainages that were clogged with massive loads of debris, ranging from clay and silt to house-sized boulders, depending on local circumstances. Given the tremendous instability of the landscapes undergoing deglaciation, it is a wonder that any type of vegetation could become established there. However, some species of plants thrive in very disturbed habitats, all the more so because they do not have to compete with the majority of plants that do not become established until the substrates settle down a little.

If you want to see the kinds of plants that do well in highly disturbed terrain, visit your nearest gravel pit or construction site. The pioneering plants taking hold on piles of sand or mounds of rubble are some of the same ones that succeeded

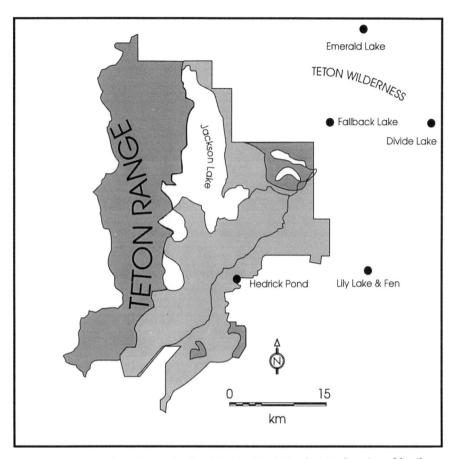

Figure 7.14. Map of Grand Teton National Park and vicinity, showing location of fossil sites discussed in the text.

very well in the topsy-turvy landscapes of the late Pinedale (and many previous deglaciation intervals). We tend to label these plants "weeds." They are a nuisance to people, because people keep giving them places in which to thrive by disturbing the soil to dig ditches, plant crops, of grow lawns.

By about 14,750 yr B.P. (i.e., almost immediately after deglaciation), Engelmann spruce was present in the vicinity of Hedrick Pond. Apparently, spruces were growing very close to the ice margins in Grant Teton National Park. However, spruce did not spread northward into Yellowstone until after 11,500 yr B.P. This northward expansion was not completed until after 10,500 yr B.P. Yet the vegetation was not a closed spruce forest. It still contained abundant meadow and shrub communities. The combination of trees and meadows formed a parkland vegetation for which there is no exact modern equivalent. Although some subalpine regions of Grand Teton and Yellowstone national parks have conifer parkland vegetation, the trees generally include pine, fir, and spruce; spruce is never the dominant tree in these communities today. Summer temperatures in this interval were probably 4–5°C (7–9°F) cooler than those today. The spread of spruce parkland in the northern Teton–southern Yellowstone region at 11,500 yr B.P. suggests increased summer temperatures and winter precipitation at that time.

Subalpine fir and pine (probably whitebark pine) were growing by Hedrick Pond by about 12,900 yr B.P., and throughout the rest of the Grand Teton–Yellowstone region by 10,500 yr B.P. Together with Engelmann spruce, they formed forests similar to the modern subalpine forests of this region. As might be expected, there were some local variations in the relative abundance of tree species. For instance, the pollen record from Fallback Lake indicates that lodgepole pine and whitebark pine were present during the interval when spruce was regionally dominant. Pine (probably whitebark pine) was present at Emerald Lake during the spruce phase, before subalpine fir pollen increased to significant levels, while at Lily Lake Fen spruce and fir pollen percentages began to rise before that of whitebark pine did.

Intuitively, it makes sense that late Pinedale forest composition would vary with local topography, substrates, and precipitation, just as it does today. Certain environmental conditions favor the establishment and growth of some tree species over others. The puzzling scenarios with which paleoecologists must come to grips are those in which some of the potential players (i.e., species of plants and animals usually associated with a given ecosystem) are absent from an ancient landscape, or those in which groups of species form communities unlike anything seen in the modern world.

The most likely explanation for the first of these scenarios (the "missing player" problem) is that the species in question was simply not in the vicinity when conditions suitable for its establishment came into being. For instance, if a given species of pine had retreated to northern Colorado during Pinedale times, it couldn't have become established in the Teton region at the end of the last glaciation because it first would have had to migrate across the intervening basin and mountain regions and then make its way north through the long process of plant migration. In the Teton-Yellowstone region, most plants, including conifers, remained nearby in unglaciated regions of Wyoming during the Pinedale Glaciation. Palynologist Whitlock estimates that Engelmann spruce migrated north into the Yellowstone region at a rate of about 200 m (660 ft) per year. It grew in unglaciated regions south of Hedrick Pond for nearly 3000 years before it colonized Grand

Teton and Yellowstone national parks. Its progress northward was not a steady advance. Rather, it developed in a patchwork fashion, spreading into localities with suitable climatic conditions and substrates.

The second scenario (the "no modern analogue" problem) poses special difficulties. If there is no modern biological community with the same makeup as an ancient community, we are forced to make educated guesses about the environmental conditions that fostered that community, based on combinations of temperature, moisture, and other conditions under which the various species of the ancient community are found living today. For instance, the addition of lots of sagebrush pollen to the late glacial vegetation that otherwise resembled modern alpine tundra makes us stop and give special consideration to the physical environments of that time. Alpine tundra meadows generally grow in fairly moist localities, but sagebrush is associated with arid landscapes. However, sagebrush *can* tolerate cold conditions, so we infer that the late glacial landscapes of the parks experienced a climate similar to that found today in the alpine zone (a cold climate), but somewhat drier than the modern alpine climate.

It is a bit whimsical, but not too farfetched, to suggest that if you want to take a walk back in time to the end of the last glaciation, you should start at the base of one of the Tetons and walk straight upslope to the top. (This is an imaginary trip, so you won't need any climbing gear, permits, or mountaineering experience!) As you climb into the subalpine forests of spruce, fir, and pine, you are leaving the Holocene and going back into late Pinedale times. Keep climbing and you will reach the boundary zone between subalpine forest and alpine tundra. Here there are conifers widely scattered in a parkland community with abundant herbs and shrubs. This community is not unlike that which developed shortly after the ice receded from the lowlands, more than three thousand feet beneath you in Jackson Hole. As you make your final ascent, you reach a forbidding landscape of bare rock, ice, and snow. This, in miniature, is the world of the last glaciation, where the rigors of the climate forced the biological world to retreat to warmer districts and ice held sway over large regions. It is a world that is hard for us to visualize. Yet, over the last two million years, this ice-bound world has been the rule, and the interglacial periods, like the one in which we live, have been the exception.

Holocene Ecosystems

Between 10,500 and 9500 yr B.P., regional pollen assemblages show marked increases in pine pollen, probably lodgepole pine. At Lily Lake Fen, Douglas-fir and aspen pollen were also important at this time. Douglas-fir was also present at Hedrick Pond and Divide Lake between 9500 and 5000 yr B.P. Lodgepole pine,

Douglas-fir, and aspen are all fire-adapted species, that is, they can withstand the effects of some fires and outcompete other trees in becoming established on fire-scarred landscapes. Regions in which these trees dominate today generally experience fires every 25–150 years, whereas regions of spruce-fir-pine forest generally have a fire frequency of about 300 years. It seems likely, then, that Grand Teton National Park, like the southern Yellowstone region, was warm and dry in the first half of the Holocene. Summer storms brought lightening strikes that started fires throughout the region, but dry environments foster the accumulation of dry tinder and dry bark, cones, and needles on conifers. These are the fuels that allow lightening-strike fires to spread across whole forests, as demonstrated in the 1988 fires in Yellowstone. From a paleo-ecological perspective, the summer of 1988 was not a disaster for the Yellowstone ecosystems. It was just a necessary stage in the ecological cycle of a fire-adapted forest.

After 5000 yr B.P., regional pollen records show an increase in lodgepole pine and a decrease in Douglas-fir. In cool, moist valleys, spruce, fir, and whitebark pine became more important elements of the vegetation. In the last 1000–2000 years, regional forests have opened up somewhat, becoming more parklike. This trend may have been due to climate change, increased fire frequency, or bark beetle attacks brought on by these changes. Bark beetles are always present in conifer forests, but their populations swell to epidemic proportions when the trees are weakened through disease, fire scarring, or drought.

Early Peoples of the Teton Region

Little is known of the first inhabitants of Grand Teton National Park. Archaeologists surmise, based on indirect evidence, that Paleoindian hunters passed through the region. Probably the first direct evidence of human activity in the park region comes from obsidian excavated from a site at Teton Pass, just south of Grand Teton National Park. This stone was made into a projectile point during the Paleoindian period and found its way north to Yellowstone National Park, so presumably the owners of the obsidian point passed through Jackson Hole on their way to Yellowstone. Conversely, Paleoindians that fashioned Clovis and Folsom projectile points from the Obsidian Cliffs site in Yellowstone transported those points to the shores of Boulder Lake, near Pinedale. The most obvious route to travel between Yellowstone and Pinedale runs through Jackson Hole, so we may be safe in assuming that early peoples passed through the park as early as 12,000 years ago. This is well after the retreat of the main Pinedale ice and into the time when conifers were becoming established. Presumably, some of the modern animal life of the park had also moved into the region. Unfortunately, we have little direct knowledge of the late Pinedale faunal history. Both paleontological and archaeological sites were covered over by the waters of Jackson Lake after the dam was built, so ancient campsites along the lakeshore are out of the reach of archaeologists. However, the water level in the reservoir is periodically drawn down to facilitate dam maintenance. During such drawdowns ancient bone beds and campsites have been found.

By 5000 yr B.P., people of the Early Archaic culture were living in the Jackson Hole region. Native populations grew in the late Holocene, as elsewhere. In Late Prehistoric times, we know that the Shoshone tribe lived in this region. Their name for the Tetons is *Teewin-ot*, or "many pinnacles."

Archaeological studies of Late Prehistoric sites in the park show that the Shoshone utilized a wide variety of game animals, including antelope, bison, elk, and mule deer. Abundant bison remains have been found in several Late Prehistoric–age sites in Jackson Hole.

John Colter, a member of the Lewis and Clark expedition, visited the Teton region in 1807 in the company of some fur trappers. Other trappers soon followed, and Jackson Hole became a dangerous place for beavers to live until the 1840s, when the fashion in men's hats shifted away from beaver skin.

Several interesting and unexpected facts have emerged from the piecing together of the ice-age history of Grand Teton National Park. Study of the geology of Jackson Hole reveals tremendous depths of unconsolidated sediments, carried down to the plains by streams of water and ice over the last two million years. The glacial geology of the region has been particularly difficult to unravel, because uppermost layers of glacial outwash in Jackson Hole, deposited at the end of the last glaciation, obscure the record of all previous events that lies underneath. Gazing up at the Tetons, one would probably guess that these mighty mountains were the source for the mantle of glacial debris that covers the plain below, but this is not the case. Most of the debris came from the larger, if less spectacular, mountain regions to the north and east. The surface area of the Tetons is simply too small to allow the development of large glaciers.

Suggested Reading

Elias, S. A. 1995. *The Ice-Age History of Alaskan National Parks.* Washington, D.C.: Smithsonian Institution Press. 150 pp.

Lageson, D. R., and Spearing, D. R. 1988. *Roadside Geology of Wyoming.* Missoula, Montana: Mountain Press. 271 pp.

Pierce, K. L., and Good, J. D. (eds.). 1992. Field Guide to the Quaternary Geology of Jackson Hole, Wyoming. U.S. Geological Survey Open File Report 92-504. 54 pp.

Whitlock, C. 1993. Postglacial vegetation and climate of Grand Teton and southern Yellowstone National Parks. *Ecological Monographs* 63:173–198.

8

ROCKY MOUNTAIN NATIONAL PARK
Life in the Rarified Air

Rocky Mountain National Park gives its visitors a good look at life at the top. The park is located in north-central Colorado (Fig. 8.1), straddling the Continental Divide (Fig. 8.2). The journey over Trail Ridge Road and the many hiking trails that extend above treeline lead visitors through hanging valleys, along cliffs, and over the ridge tops of mountains sculpted by glacial ice (Fig. 8.3). The park contains a segment of the Continental Divide as spectacular as any in the central Rocky Mountains, with more than sixty peaks rising above treeline (about 3350 m, or 11,000 ft).

The land above the trees has more in common with the high arctic than with the forests that surround it. Tundra plants are all that can survive here, enduring the short growing season, howling winds, and frigid temperatures. They can make do with the moisture that comes their way, which is surprisingly little. Snow may fall in the alpine as heavily as elsewhere in the mountains, but little of it remains on the bare, windswept slopes above treeline, except in the hollows and stream courses, where the winds diminish. This is a special world, a fragile world that is easily disturbed by human activity. The mountaintops are the last refuge in this region for a tundra biota that fared far better during the Pleistocene glaciations.

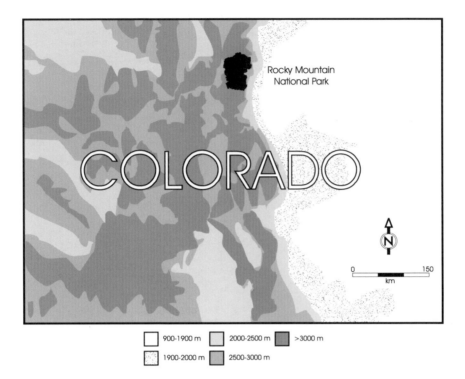

Figure 8.1. Map of Colorado, showing location of Rocky Mountain National Park and elevation regions.

The climate of the alpine zone is as close to an arctic climate as you will see in the "lower 48" states. Summer daytime temperatures rarely exceed 15°C (60°F), and the temperature falls below freezing on many summer nights. Winter temperatures remain below freezing for many months, and nighttime lows may plummet to −40°C (−40°F). Patches of permafrost (permanently frozen ground) in the alpine produce such features as **frost boils** and **sorted stone polygons** (Fig. 8.4)

The weather in the alpine can change rapidly from summer to winter conditions. I have been caught in blizzards on Trail Ridge Road in July, while the eastern plains of the Front Range region were basking in sunshine. On other summer days, I have stretched out on a carpet of alpine tundra meadow, enjoying the warmth of the sun on the mountaintop while the valleys below were swathed in rain clouds. All but the most physically fit soon learn not to run around too much in the alpine. The air here is quite a bit thinner than that at sea level—as your lungs will tell you when you gasp for air. In many ways, the alpine is a world unto itself, and one that

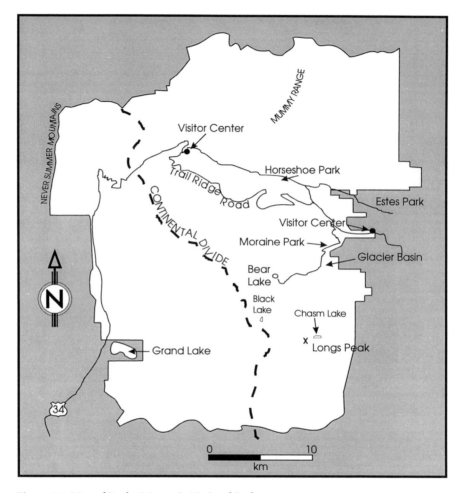

Figure 8.2. Map of Rocky Mountain National Park.

must be met on its own terms. As people of the industrial twentieth century, we are used to dominating landscapes and having our own way with the environment, but no one has truly tamed the alpine. I hope it remains that way.

Rocky Mountain National Park contains big patches of alpine landscape: islands of tundra in a sea of forest. There was a time, however, when those islands were a lot bigger; they crept downslope and came together to form a peninsula of alpine environments. Tongues of ice outgrew the cirques where they began and spilled out of the high mountains into the valleys below. This chapter provides a summary of these events and of how they affected the plants and animals of the region as well as the first humans to enter the land.

Figure 8.3. View looking west along the Continental Divide from Trail Ridge Road. Note the hanging valleys, cirque (Hayden Gorge), and arête (Hayden Spire), all glacial features. (Photograph by the author.)

Modern Setting

The northwest corner of the park is bounded by the Never Summer Range, the northeast corner by the Mummy Range. The Front Range forms the Continental Divide through much of the park, capped by Longs Peak in the southeast corner of the park, soaring above its neighbors to a height of 4350 m (14,255 ft). The Precambrian rocks that form the summit of Longs Peak are 6700 m (22,000 ft) higher than the Precambrian rocks in the deepest part of the Denver Basin. The displacement between peaks and plains was probably even greater in the past, as the mountaintops have since been worn down by erosion.

The ecosystems of the park are outlined in Figure 8.5. The lowest regions on the east side of the park (near the town of Estes Park) are in ponderosa pine parkland. Douglas-fir dominates north-facing slopes in the lowlands east of the Continental Divide and is more widespread in the lowlands west of the divide, which receive more moisture than the east. Douglas-fir and lodgepole pine dominate the forest stands above 2590 m (8500 ft) in the upper montane forest.

Above 2750 m (9020 ft), these montane forests give way to subalpine forests of Engelmann spruce and subalpine fir. Limber pine grows on rocky slopes and

Figure 8.4. Patterned ground of stone nets, alpine tundra, Trail Ridge Road, Rocky Mountain National Park. (Photograph by the author.)

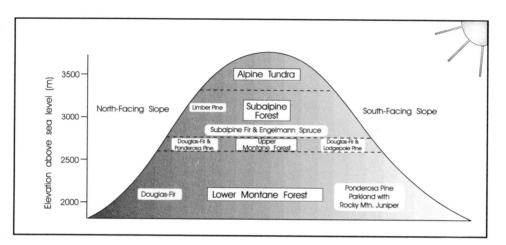

Figure 8.5. Modern ecosystems of Rocky Mountain National Park. (Data from Marr, 1967.)

windy promontories. In the lower parts of the subalpine, spruce and fir grow in tall, dense stands. Near the upper limit of the forest, however, the trees get shorter and shorter, taking on dwarf, twisted **krummholz** forms at or near treeline (Fig. 8.6). Krummholz trees only a few tens of centimeters (a few inches) tall creep out into the alpine tundra, starting around 3350 m (11,000 ft). The boundary (or *ecotone*) between subalpine and alpine ecosystems is therefore not a sharp, well-defined line of trees at the edge of the tundra. Rather, it is a mosaic of open ground and conifer shrubs. By about 3475 m (11,400 ft), the krummholz "tree islands" disappear in a sea of tundra, rock, and ice.

The large mammals of the park include black bear, elk, mule deer, bighorn sheep, coyotes, and mountain lions. Elk and deer had nearly been eliminated by overhunting in the Colorado Front Range by the beginning of the twentieth century, but both have made spectacular comebacks in recent years. Moose, never very abundant here at the southern edge of its range, has recently been reintroduced into the park and adjacent regions of the Rockies in northern Colorado. Grizzly bears have not been seen in this region for many years.

Late Pleistocene Glaciations

The history of glaciation is not as well worked out for the Rocky Mountain National Park region as it is for Grand Teton and Yellowstone national parks. For

Figure 8.6. Krummholz tree island on alpine tundra, Trail Ridge Road, Rocky Mountain National Park. (Photograph by the author.)

instance, although a chronology of three separate ice advances has been established for Grand Teton National Park during Pinedale times, in northern Colorado we know only that there were earlier and later Pinedale ice advances. We do not know when the earlier advance (or multiple advances) took place. However, based on geologic evidence compiled by Richard Madole and colleagues at the U.S. Geological Survey, the early Pinedale glaciation was more extensive than the late Pinedale. Early Pinedale moraines can be seen near the east entrance to Rocky Mountain National Park, whereas late Pinedale ice formed moraines several miles up-valley.

In general, the Bull Lake Glaciation (the last glaciation before the Sangamon interglacial period) left end moraines a kilometer or two farther down-valley than Pinedale ice advanced (Fig. 8.7). Bull Lake and Pinedale glaciers flowed out of high mountain cirques, down-valley to elevations between 2440 and 2470 m (8000–8100 ft), the elevation of the lower montane forests. Pinedale ice advanced 14–15 km (9–10 miles) downslope from the Continental Divide on the eastern slope. The western slope received more moisture in Pinedale times, as it does today. So the Pinedale glaciers on the western slope were larger, extending downslope as much as 33 km (20 miles). The largest glacier formed in the park was the Colorado River Glacier, which flowed from the alpine headwaters of the drainage, merging with ice from cirques in the Never Summer Range to form one large river of ice

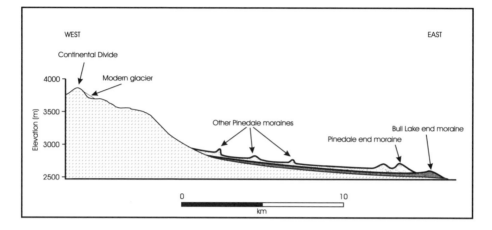

Figure 8.7. Limits of Bull Lake and Pinedale till and moraines on the eastern slope of the Colorado Front Range (vertical scale of till and moraines exaggerated). (Data from Madole and Shroba, 1979.)

that crept 33 km down the Colorado River drainage. Pinedale ice may have been as much as 450 m (1500 ft) thick near the heads of the glaciers in the park.

We have good estimates of the timing of some Pinedale events, based on radiocarbon ages of organic-rich sediments at several high-elevation sites. At the Mary Jane ski area, near Winter Park, Colorado (Fig. 8.8), an excavation for a ski lift tower exposed a series of alternating lake sediments and **glacial tills.** According to U.S. Geological Survey geologist Alan Nelson and colleagues, the oldest lake bed was dated at about 30,000 yr B.P. This bed was overlain by glacial tills, and the next youngest lake bed yielded a radiocarbon age at the base of about 13,750 yr B.P. Based on the Mary Jane sequence, it appears that the last major ice advance of the Pinedale Glaciation took place between the time of deposition of the older and younger lake beds, that is, between 30,000 and 13,750 yr B.P.

At Devlins Park, south of Rocky Mountain National Park on the eastern slope of the Front Range (Fig. 8.8), geologists Thomas Legg and Richard Baker studied sediments from a lake that was dammed by ice in the late Pinedale. During the time that Glacial Lake Devlin existed (22,400–12,200 yr B.P.), ice was residing in the Devlins Park region. The lake drained catastrophically when the ice retreated, and sediments from the top of the lake bed profile provide an age for this event. Devlins Park is at an elevation of 2953 m (9690 ft). Presumably the late Pinedale glacier that advanced downslope from the Continental Divide region west of Devlins Park in the Indian Peaks Wilderness area took

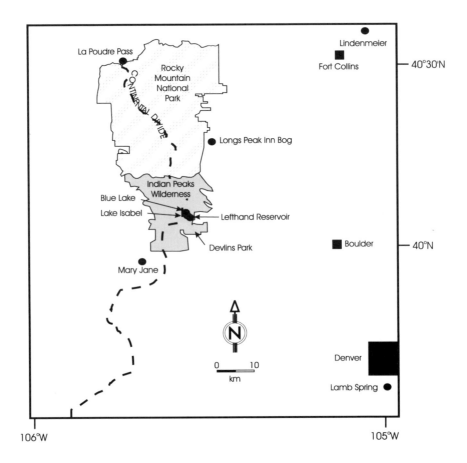

Figure 8.8. Map of the Front Range region of Colorado, showing location of fossil and archaeological sites discussed in the text.

some centuries to reach this elevation, so the glacial advance began before 22,400 yr B.P. The terminal moraine for this glacier is 2.3 km (1.4 miles) downslope from the site of the lake.

We can trace the retreat of late Pinedale ice from montane valleys back to the alpine zone where they originated. At La Poudre Pass, on the northern boundary of Rocky Mountain National Park (Figs. 8.8 and 8.9), U.S. Geological Survey geologist Richard Madole obtained radiocarbon ages from the base of a peat section (the peat formed in a bog after Pinedale ice retreated). These samples have been dated at about 10,000 yr B.P., indicating that the pass, which is located at modern treeline, was free of ice sometime prior to that time.

Figure 8.9. The author at the La Poudre Pass fossil site. A peat bank was exposed here by the excavation of an irrigation ditch. (Photograph by Harvey Nichols, University of Colorado.)

Glacial Features Easily Viewed in the Park

Pinedale moraines are still clearly visible in many locations, especially the **lateral moraines.** These are sharp-crested, steep-sided ridges, from a few meters to many tens of meters (a few feet to more than 100 ft) high. A good example of these are the long ridges that enclose Moraine Park (Fig. 8.10). Although lateral moraines are usually prominent features, end or terminal moraines are not often conspicuous, as they may have been buried by subsequent outwash or partially removed by stream erosion. Older glaciations (pre-Pinedale and pre–Bull Lake) have left fewer obvious signs on the landscape. However, sometime earlier in the Pleistocene, a glacier advanced out of the mountains in the park, filling the Estes Park Valley and terminating near the head of St. Vrain Canyon, below the town of Estes Park. Just below the park's main entrance station is a terminal moraine from the Bull Lake Glaciation. About 1 km (0.6 mile) above the station lies the outermost terminal moraine of Pinedale age. This moraine has a sandy soil, and fresh-looking boulders sit on it.

As you climb Trail Ridge Road, you look down on Horseshoe Park and Moraine Park from a narrow ridge between these two broad, glaciated valleys. Horseshoe

Figure 8.10. Moraine Park, showing lateral moraines of the Pinedale Glaciation (arrows). (Photograph by the author.)

Park has lateral and terminal moraines from a glacier that flowed down the Fall River drainage in Pinedale times. The glacial features of Moraine Park were formed by ice from the Thompson Glacier, the glacier of the Big Thompson River drainage.

Most of the lakes in Rocky Mountain National Park are the product of glaciation. Several of the park's lakes occupy kettle holes formed as stagnant ice melted during Pinedale deglaciation. Among these are Bear Lake (Fig. 8.11), Bierstadt Lake, Copeland Lake, Dream Lake, and Sheep Lake. These lakes formed *on* moraines, so their shores are littered with boulders and other glacial debris. Other lakes in the park are *tarns*, lakes that form at the bottom of cirques. When water fills the depression within the cirque, a tarn is formed. Among the tarns in the park are Blue Lake, Black Lake, Chasm Lake, Fern Lake, Odessa Lake, Shelf Lake, and Tourmaline Lake. A third form of glacial lake is the *pater noster* lake, so named because of the resemblance of a group of these lakes to a string of prayer beads. Pater noster lakes form when a glacier advancing across a surface encounters

Figure 8.11. Bear Lake, a kettle lake formed in a depression left by stagnant ice at the end of the Pinedale Glaciation. (Photograph by the author.)

"weak zones," areas where the bedrock is more easily eroded. The glacier gouges out material from these zones. The series of depressions later fills with water, forming a chain of lakes, drained by a single stream (Fig. 8.12). Cub Lake, East Inlet Lake (including Lake Verna), Loch Vale (Fig. 8.13), and Ouzel Creek Lake are all pater noster lakes.

Glacial erratic boulders are boulders that have been transported by glacial ice to locations down the ice stream from their source. There are several localities in the park where glacial erratic boulders may be seen. One of these is called the Boulder Field and another is along the banks of Fern Creek.

As you climb farther up the road to Trail Ridge, you view Fall River in Horseshoe Park, but for more than a mile Horseshoe Park is hidden from view by a long lateral moraine of Pinedale age. This moraine also conceals Hidden Valley. As the road climbs higher into the subalpine and alpine zones, many glacial features can be seen. The cliffs, cirques, and hanging valleys of the uplands were all scoured by glacial ice.

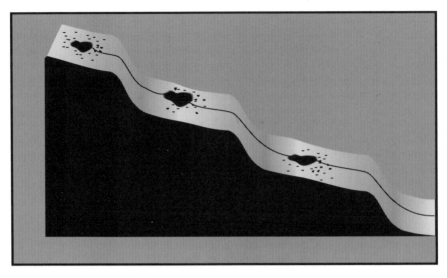

A

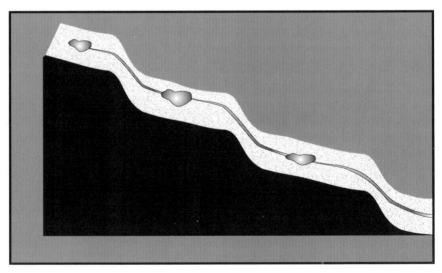

B

Figure 8.12. Development of pater noster lakes in a glaciated high-elevation landscape. (A) Glacial ice leaves scoured pits and basins at points of weakness in bedrock surface. (B) Scoured basins fill with water to form a string of pater noster lakes drained by one stream.

Figure 8.13. Pater noster lakes in the Loch Vale watershed, Rocky Mountain National Park: below the Taylor Glacier lie Sky Pond and Glass Lake. (Photograph by Jill Baron, Colorado State University.)

Figure 8.14. View from the visitor center at Fall River Pass, showing a glacial cirque (arrow). (Photograph by the author.)

From the Forest Canyon overlook, look down 750 m (2500 ft) into the canyon. This is a steep-walled trough, gouged out by glacial ice. Before the Pleistocene glaciations, the Big Thompson River may have followed a winding course through this region, but the shape of the valley was altered by repeated advances of ice. Glaciers don't like to turn corners, so they set about straightening the valleys that they flow through. Across the canyon are a series of small lakes in **hanging valleys,** cut by glaciers, that climb upslope in stairstep fashion. These lakes fill depressions scoured out by glacial ice.

At Fall River Pass, look down to see clusters of cirques at the head of Fall River Canyon (Fig. 8.14). On the west side of the pass at Farview Curve, look down into the Kawuneeche Valley. The Colorado River now meanders through the long, straight valley that was gouged out by Pleistocene glaciers. Grand Lake, at the western entrance to the park, is a natural lake that was dammed by two moraines. One of these is a lateral moraine of a glacier that once filled Kawuneeche Valley; the other is the terminal moraine of a glacier that came down the Paradise Creek drainage.

Shifting Life Zones in the Late Pleistocene

So far, we have only sketchy reconstructions of late Pinedale environments in and around Rocky Mountain National Park. The earliest postglacial site in the region is the peat bog at La Poudre Pass (Fig. 8.8), with a basal age of about 10,000 yr B.P. Older sites in the mountain region are few and very far between, possibly because they were scoured away by glacial ice in the late Pinedale.

The oldest Pinedale site in the region is the Mary Jane site, near Winter Park. Here pollen in lake sediments laid down during an interstadial interval before the last major Pinedale ice advance (circa 30,000 yr B.P.) records a sequence of vegetation beginning with open spruce-fir forest with herbs and shrubs, adjacent to the lake. This is followed by a colder phase, in which alpine tundra replaces the subalpine forest. The youngest (uppermost) sediments in this lake bed reflect the return of spruce forest to the vicinity before the advance of late Pinedale ice. The Mary Jane site is at an elevation of 2882 m (9450 ft), in the lower part of the modern subalpine forest. The existence of alpine tundra at the site in mid-Pinedale times translates into a depression of treeline by more than 500 m (1640 ft). This, in turn, corresponds to a climatic cooling of at least 3°C (5°F) from average modern summer temperatures.

The next indication of late Pinedale environments in the region comes from the oldest lake sediments from Glacial Lake Devlin (22,400 yr B.P.). Legg and Baker found pollen evidence for a treeless landscape at the site. The pollen diagram is dominated by sagebrush. Scattered pollen grains of tundra plants (which are not abundant pollen producers) and low numbers of conifer pollen grains suggest that the site was above treeline in the late Pinedale. The pollen assemblages from this site are in many ways similar to those described by Baker and Whitlock from the earliest postglacial environments in Grand Teton and Yellowstone national parks. In both instances, mixtures of sagebrush and alpine tundra plants suggest steppe-tundra. This vegetation probably developed in the sort of cold, dry climate seen today in parts of Siberia.

We pick up the paleoenvironmental story again at the Mary Jane site, where peat layers were deposited after the retreat of late Pinedale ice. Palynologist Susan Short and I studied these peats, which range in age from 13,740 to 12,350 yr B.P. Insect fossil assemblages, plant macrofossils, and pollen from 13,740 through 12,700 yr B.P. suggest open ground environments with flora and fauna of the alpine tundra. Summer temperatures were as much as 5–6°C (9–11°F) cooler than today during this interval. After 12,600 yr B.P., montane and subalpine insects increased at the expense of alpine species, and willow shrubs expanded across the bog, suggesting a climatic warming. By 12,300 yr B.P., spruce trees were probably growing adjacent to the site, after an absence of about 18,000 years.

An archaeological site called Lamb Spring, in the foothills just east of the Front Range (Fig. 8.8), has yielded insect assemblages that provide information on late Pinedale environments of lower elevations. This site was excavated by Smithsonian archaeologist Dennis Stanford in the 1970s. A faunal assemblage radiocarbon dated at 17,850 yr B.P. includes species found today in cold, dry grassland regions as far north as Alberta. A younger assemblage, dated at 14,500 yr B.P., is indicative of cold, wet conditions. Some of the species in this younger assemblage are found today in the alpine tundra zone of Colorado. The assemblages came from a spring deposit that also contained the bones of mammoth, American camel, and other Pleistocene megafauna.

Ice-Age Hunters of the Central Rockies and Their Prey

The earliest humans to populate the highlands of the central Rocky Mountains were probably people of the Clovis culture, the earliest documented Paleoindians in the western United States. The name *Clovis* comes from the first site where artifacts of this culture (projectile points) were found, near Clovis, New Mexico. The earliest radiocarbon age associated with Clovis artifacts is about 11,500 yr B.P. Clovis peoples traveled widely throughout the Rocky Mountain region, leaving but a few stone tools as evidence of their passing. They were apparently nomadic hunters of big game. At several sites in the Rocky Mountain region, Clovis artifacts have been found associated with mammoth bones. We know that Clovis hunters visited the Rocky Mountain National Park region at least occasionally, because a Clovis projectile point was found near treeline at the east end of Trail Ridge. Based on extensive surveys of the Continental Divide regions in the park and farther south in the Indian Peaks Wilderness, archaeologist James Benedict has suggested that these early Paleoindians spent little time in the high country of Colorado. There are Clovis-age artifacts in this region, but nearly all of them have been found either on mountain passes or in valleys leading to passes. Thus it appears that these early hunters crossed over the mountains, but spent little time hunting there.

The original discovery of artifacts of the so-called Folsom people, also in New Mexico, proved that they also lived in the Pleistocene, because a Folsom point was found embedded in a Pleistocene bison (*Bison antiquus*) bone. The earliest Folsom site in northern Colorado is the Lindenmeier site (Fig. 8.8), east of Rocky Mountain National Park in the piedmont of the Front Range. At this site, Folsom fire hearths from a camp have been radiocarbon dated to about 10,900 yr B.P. That age is within 200 years of the age of the original Folsom site in New Mexico. The artifacts from the site are more varied than those from many other Folsom sites. Besides projectile points, the Lindenmeier site also contained scrapers, knives, engraved pieces of bone (including pieces that may have been used in games), and bone needles. People at this site had killed and eaten pronghorn antelope, rabbit, fox, wolf, coyote, turtle, and bison. Thus we know that Folsom peoples were hunting on the eastern plains of Colorado at that time, but what about the high country?

There is good evidence that Folsom hunters were making more use of the high country by the end of the Pinedale (about 10,500–10,000 yr B.P.). The Pleistocene megafauna was gone by then, but other game animals, including elk and bighorn sheep, would have been worth pursuing. These animals are frequent visitors to the alpine, and Paleoindians erected game drives to help trap them. Today evidence of these game drives remains on many high slopes of the Front Range as piles of rocks forming a line across a ridge or high valley. When they were built, sticks or brush were probably added to the stone walls, providing more camouflage for the hunters lying in wait for the animals. In some places, corrals were built at the ends of the game drive "funnel." In other places, the animals were directed off a cliff at the end of the game drive walls. The effectiveness of game drives as a hunting technique is attested to by the number of game drives still visible and the obviously considerable work that went into their construction.

Holocene Environments: Islands of Tundra in Seas of Forest

During the Holocene, Rocky Mountain National Park and its environs experienced a series of climatic fluctuations. Insect assemblages indicative of warmer-than-present conditions have been identified from La Poudre Pass and dated between 9500 and 7000 yr B.P. Conditions colder than those at present between 4500 and 3000 yr B.P. are indicated by insect faunal assemblages from La Poudre Pass, Lake Isabel, and Lefthand Reservoir. The latter two sites are in the Indian Peaks Wilderness area, south of the park. Climatic cooling between 2000 and 1000 yr B.P. is suggested by fossil insect data from these sites and from a bog near the eastern foot of Longs Peak. During this interval, the beetles in the fossil assemblages are indicative of upper subalpine and alpine tundra environments. Today the site is in the upper montane forest.

Insect responses to climate change in this region have occurred at essentially the same time as changes in vegetation inferred from pollen assemblages. However, the insect data suggest that conditions at the beginning of the Holocene were warm enough to allow spruce-fir forest to grow up to its modern limit. Pollen records studied by Susan Short indicate that subalpine forests did not reach modern treeline until 9500–9000 yr B.P. Thus it appears that the trees lagged behind early Holocene climate change by 500–1000 years.

Whatever the pace of change, the end result was that the broad belt of alpine tundra that developed in Pinedale times at elevations down to about 2900 m (9500 ft) retreated upslope to the mountaintops. The tundra peninsula became an archipelago of tundra islands. We do not know very much about the composition of alpine tundra vegetation during the Pinedale Glaciation for two reasons. First, most tundra plants do not produce much pollen, so they are poorly represented in Pinedale-age pollen assemblages, even those that come from high-elevation sites. Second, it is difficult if not impossible to identify most pollen grains to the species level, so we know only the genera and not the species of most plants in pollen assemblages.

We know more about changes in fauna than changes in flora. Many species of cold-adapted animals that flourished in the Rockies during the Pinedale either are now extinct or have been extirpated (that is, they are still alive, but no longer live in the region). The extinct animals include American camel and horse, Pleistocene bison, woodland musk-ox, mammoth, and mastodon. Extirpated animals include caribou, musk-ox, and lemming.

We also know a bit more about what happened to the Pinedale insect fauna of the Rockies. The cold-adapted insects that lived at Lamb Spring, Mary Jane, and some other late Pinedale–age sites had two options at the end of the glaciation. They could either resettle in the ever-shrinking alpine tundra on the tops of mountains or head north, leaving this region behind. Both strategies were adopted. Some species that were found far below the mountaintops in the Pinedale Glaciation now live in the alpine tundra of Rocky Mountain National Park and adjacent tundra "islands." Other species apparently have not lived in the Rocky Mountain region since the late Pinedale. These now live in arctic and subarctic regions of Canada and Alaska. However, given that the vast majority of time over the last two million years has been spent in glaciations, with relatively brief warm intervals in between long glacial intervals, it seems very likely that some of these arctic creatures will return again. They are simply waiting out the current interglacial, as their ancestors did many times before.

The vegetation history of Rocky Mountain National Park and adjacent Indian Peaks Wilderness has been interpreted from fossil pollen data by palynologist Susan Short. One of the study sites is Long Lake, in the Indian Peaks Wilderness (Figs. 8.8 and 8.15). The lake lies near treeline in the upper subalpine zone. At this site, pollen from the lake sediments shows that alpine tundra vegetation became established after deglaciation and persisted from 12,000 yr B.P. to 10,500 yr B.P., when spruce and fir arrived. However, the presence of these conifers does not mean that subalpine forest became established at that time, because elements of the alpine tundra persisted there until about 9500 yr B.P. This trend is also seen from data collected at La Poudre Pass, where a mixture of alpine tundra and spruce

Figure 8.15. Researchers taking a core from Long Lake, Indian Peaks Wilderness, Colorado. (Photograph by Susan K. Short.)

woodland developed after deglaciation, from 9800 to 9100 yr B.P. Sagebrush, grasses, and other herbs dominated the ground cover. From about 9000 to 6800 yr B.P., spruce-fir forest and some pine (probably lodgepole pine) grew at La Poudre Pass and Long Lake, but treeline was still probably below its modern elevation at the former site. The vicinity of the bog at La Poudre Pass still had much open ground with herbaceous cover, but alpine tundra vegetation had retreated upslope by this time.

The time of maximum Holocene warmth, as indicated by pollen assemblages from La Poudre Pass, Long Lake, and other sites, came between 6500 and 3500 yr B.P. During this interval, pine expanded upslope into the modern subalpine zone throughout the Front Range. The upper limit of spruce-fir forest may have crept higher during this interval as well, but the evidence for this conclusion is unclear at the sites.

During the last 3500 years, pine retreated back to the elevations of the modern montane forests, and levels of herbaceous pollen increased once again. During this interval, the modern pattern of regional vegetation became established. The cooling associated with the beginning of this phase may have begun as early as 4000 yr B.P., as indicated by the pollen record from Blue Lake, a high-altitude (3450 m; 11,320 ft) lake in the Indian Peaks Wilderness area (Fig. 8.8).

Thus the modern ecosystems of Rocky Mountain National Park are not immutable associations of plants and animals that have been locked together for many thousands of years. Instead, here as elsewhere, the modern ecosystems represent merely the latest reshuffling of species. Think of the park as an ecological stage on which a biological play is being acted out. There have been many acts in this play, even since the last deglaciation. Few really new characters have been added to the script for a long, long time, because all the old players keep hanging around, waiting in the wings for their big chance to dominate a scene. New combinations of existing players come together at regular intervals. Ecological bit players are occasionally given leading roles, while some of the stars of yesteryear are forced to play smaller parts for the time being. None of the players wants to exit the stage we call Rocky Mountain National Park. The Continental Divide forms such a nice backdrop, and the stage settings are truly beautiful.

Suggested Reading

Cassells, E. S. 1983. *The Archaeology of Colorado*. Boulder, Colorado: Johnson. 325 pp.

Chronic, H. 1980. *Roadside Geology of Colorado*. Missoula, Montana: Mountain Press. 335 pp.

Elias, S. A. 1985. Paleoenvironmental interpretations of Holocene insect fossil assemblages from four high-altitude sites in the Front Range, Colorado, U.S.A. *Arctic and Alpine Research* 17:31–48.

Elias, S. A. 1991. Insects and climate change: Fossil evidence from the Rocky Mountains. *BioScience* 41:552–559.

Harris, A. G. 1977. Rocky Mountain National Park. In *Geology of the National Parks*. 2d ed. Dubuque, Iowa: Kendall/Hunt, pp. 85–94.

Legg, T. E., and Baker, R. G. 1980. Palynology of Pinedale sediments, Devlins Park, Boulder County, Colorado. *Arctic and Alpine Research* 12:319–333.

Madole, R. F., and Shroba, R. R. 1979. Till sequence and soil development in the North St. Vrain drainage basin, east slope, Front Range, Colorado. In Ethridge, F. G. (ed.), *Guidebook for Postmeeting Field Trips Held in Conjunction with the 32nd Annual Meeting of the Rocky Mountain Section of the Geological Society of America, May 26–27, 1979, Colorado State University*. Fort Collins, Colorado: Geological Society of America, pp. 124–178.

Marr, J. W. 1967. Ecosystems of the East Slope of the Front Range in Colorado. University of Colorado Studies, Series in Biology No. 8. 134 pp.

Nelson, A. R., Millington, A. C., Andrews, J. T., and Nichols, H. 1979. Radiocarbon-dated upper Pleistocene glacial sequence, Fraser Valley, Colorado Front Range. *Geology* 7:410–414.

Short, S. K. 1985. Palynology of Holocene sediments, Colorado Front Range: Vegetation and treeline changes in the subalpine forest. *American Association of Stratigraphic Palynologists Contribution Series* 16:7–30.

Short, S. K., and Elias, S. A. 1987. New pollen and beetle analysis at the Mary Jane site, Colorado: Evidence for Late-Glacial tundra conditions. *Geological Society of America Bulletin* 98:540–548.

9

CONCLUSION

What Can We Learn from the Past?

My hope in writing this book has been that you will catch some of the sense of wonder and fascination for the Rocky Mountain wilderness, as preserved in its national parks. But more than that, I hope that you have come to understand that what can be seen there today, however awe-inspiring, is only the latest page in the very long book of life written for this great region. That book is a strange and unfamiliar one to most people, since it is written in peat, silt, and clay. Because it is unfamiliar, most have overlooked it. But once you have been introduced to it, the ancient past has a lot to offer.

From this historical perspective (some might call it the prehistoric perspective), you begin to see nature in a new way, by gaining an appreciation for the fact that modern ecosystems and landscapes have been shaped by past events. As I mentioned at the end of the last chapter, the current crop of plants and animals in biological communities are just the most recent biological actors to appear on stage for one act of a very long play. Furthermore, the combination of species that occurs in a specific ecosystem may or may not be the one best suited for that particular environment. Some will probably be gone in a few centuries. Newcomers, better adapted to the system, will squeeze them out. The scene appears stable to our eyes, but in reality it is constantly shifting. So when you come to the Rockies and stand on a high promontory, look over the landscape and ask yourself, "How did it come to be this way?"

The Fragility of the Parks' Modern Ecosystems

Glacier, Yellowstone, Grand Teton, and Rocky Mountain national parks preserve large patches of wilderness. The National Park Service has been given a dual mandate that often seems contradictory: to preserve these wilderness areas and at the same time to provide public access to them. How can you preserve something as vulnerable as a mountain ecosystem while you facilitate its invasion by millions upon millions of visitors? One of the chief ways of overcoming this hurdle is to teach people to tread softly in the wilderness, to treat it with respect, and to give it the consideration it deserves. After all, it is the wilderness—that sense of wild, unspoiled nature—that endears these parks to their visitors. Let's face it: you can have a picnic or set up your campsite wherever there is a grassy spot, water, shade, and level ground. But in how many places can you hear an elk bugling or watch a mama black bear steer her twin cubs up a tree?

We are always in danger of loving these parks to death. This happens when we treat the wilderness as just another theme park or tourist attraction, rather than as the precious, rare commodity it really is. Having visited Europe, where there is essentially no wilderness left at all (that which appears to be wilderness has actually been shaped and managed by people for many centuries), I appreciate our Rocky Mountain wilderness regions all the more. I hope that the descriptions in this book of how things came to be the way they are will add to your sense of wonder and your appreciation of these magnificent landscapes. But the bottom line is this: we must be careful with them, or they won't last for very long.

Where Do We Go from Here?

I have described in this book how the ecosystems of the Rocky Mountains follow cycles in response to climate change brought on by glacial and interglacial periods. This is true up to a point, but it must be emphasized that the current interglacial is different from the many that preceded it, because of the presence of mankind.

The growing human population is using up the world's resources at an alarming rate. We've reached the steep part of the growth curve, the part where population doubles every few decades. In the western states, we talk about "sustainable growth," but this term really contradicts itself. No growth of population can be sustained indefinitely. We have now managed to alter the natural world to the point at which natural biological cycles have been seriously interrupted. In the ancient past, these cycles tracked the course of climate change. The Earth's biota may not be able to make the necessary responses as the climate continues to change, now that people have all but eliminated the natural order in so many regions.

We are indeed fortunate that, in the Rocky Mountain region, the elements of the primeval ecosystems have remained intact, at least in fairly large patches here and there. Let us do all that we can to preserve as much of the Rockies in as natural a state as possible. This conservation ethic ranges from letting wolves back into Yellowstone to preserving old-growth forests so that their inhabitants will have a place to live. We need to take stock of the situation from a truly long-term perspective. It will show us that, like the Paleoindians, we are just passing through the Rockies. We may enjoy them for a season, but they are the product of thousands of years of environmental change, preceded by millions of years of mountain building and evolution. We do not want to be known as the people who brought an end to the wilderness in the Rockies, but because of our technology and the size of our populations, we have the capability of ruining what remains of the wilderness. This does not have to happen, but we will have to work hard to prevent it from happening.

GLOSSARY

Acidic sediments Sediments that contain more hydrogen ions than hydroxyl ions; sediments with a pH less than 7.0. Examples include some organic-rich soils and peats.

Alkaline sediments Sediments that contain more hydroxyl ions than hydrogen ions; sediments with a pH greater than 7.0. Examples include marls or soils rich in calcium carbonate.

Alpha (α) particle A helium nucleus, given off by the nuclei of certain radioactive elements.

Bering Land Bridge The continental shelf regions between Alaska and Siberia that remained above sea level for large parts of the Pleistocene.

Beringian Steppe-Tundra A Pleistocene ecosystem that occupied regions extending from western Europe across Asia and the Bering Land Bridge to Alaska and the Yukon Territory. A mixture of arctic tundra and Asian steppe vegetation, it was pervasive during glacial intervals.

Beta (β) particle High-energy electron given off by radioactive decay.

Biota All the species of animals and plants occurring within a certain region.

Block faulting A type of normal faulting in which the Earth's crust is divided into fault blocks of different elevations and orientations. It is the process by which block mountains are formed. See **fault** and **normal fault**.

Catchment basin   A basin that accumulates the runoff of precipitation from a watershed.

Chitin   A nitrogen-containing polysaccharide (carbohydrate) compound that forms the hard, outer layer in the skeletons of insects and other invertebrates.

Cirque   A steep-walled, half-bowl–shaped hollow, situated high on the side of a mountain; produced by the erosion from a mountain glacier.

Cordilleran Ice Sheet   A Wisconsin glaciation ice sheet that covered most of western Canada, extending south into Washington, Idaho, and Montana.

Crustose lichen   A flat, disk-shaped lichen that grows close to a rock or other surface.

Cyclotron   An electromagnetic machine that accelerates high-energy particles (e.g., protons and electrons) in a circular path; the particles approach the speed of light.

Diatom   A single-celled, microscopic alga characterized by cell walls reinforced with silica.

Ecosystem   A biological community, including all its component plants, animals, and other organisms, together with the physical environment, forming an interacting system.

Ecological succession   The replacement of one kind of biological community by another; the progressive changes in a region's flora and fauna that may culminate in a stable, climax community.

Exoskeleton   The external skeleton of insects.

Fault   A fracture or zone of fractures in the Earth's crust in which there has been movement parallel to the break.

Fellfield   An open, treeless, rock-strewn area above treeline or in a high-latitude region.

Fluvial sediments   Sediments laid down by running water (streams and rivers of all sizes).

Frost boil   A low mound developed by local differential frost heaving at a place most favorable for the formation of segregated ice and characterized by an absence of insulating vegetation cover.

Glacial erratic   A boulder gouged out of bedrock by glacial ice, carried along with the ice flow, and eventually dropped as the ice recedes.

Glacial moraine   A mound or ridge of unsorted glacial debris, deposited by glacial ice in a variety of landforms.

Glacial outwash   Sediments deposited by streams emanating from the fronts of glaciers.

Glacial till   Material laid down directly by glacial ice.

Hanging valley   A glacial valley whose mouth is at a relatively high level on the steep side of a larger glacial valley. The larger valley was scoured out by the main body of a glacier (the trunk glacier); the hanging valley above it was scoured by a smaller, tributary glacier. The trunk glacier eroded the large valley more deeply than did the tributary glacier, so the upper valley was left at the higher elevation.

Holocene epoch   An epoch of the Quaternary Period, spanning the interval after the last glaciation (10,000 yr B.P. to recent).

Interglacial   A long interval between glaciations in which the climate warms to at least the present level.

Interstadial   A relatively warm climatic episode during a glaciation, marked by a temporary retreat of ice.

Ion   An electrically charged atom or group of atoms. An atom with a high affinity for electrons may acquire an electron, thus becoming negatively charged.

Isotope   A variety of an element. Isotopes of an element differ from one another in the number of neutrons contained in the atom's nucleus.

Krummholz   A growth form of trees, frequently found in stands growing near treeline. The trunks of the trees are shortened and twisted; their branches lie prostrate, near the ground.

Lateral moraine   A long, low ridge deposited at or near the side margin of a mountain glacier.

Laurentide Ice Sheet   An ice sheet covering most of eastern and central Canada and the northeastern and north-central United States during the Wisconsin glaciation.

Macrofossil   The macroscopic (easily seen by the naked eye) remains of ancient organisms.

Marl   A sediment deposited in lakes, composed mainly of calcium carbonate, and mixed with clay or silt.

Megafaunal mammals   Mammal species whose adults have a live weight greater than 40 kg (88 lb).

Microfossil   The microscopic (poorly seen by or invisible to the naked eye) remains of an ancient organism.

Microsculpture   Microscopic sculpture—including striations, punctures, and meshes—on the surface of such objects as insect exoskeletons, mollusk shells, and seed coats.

Normal fault   A break in the Earth's crust in which one member has moved downward relative to the other, rather than sideways.

Nunatak   A mountain region, thought to have been above the level of glacial ice, that may have supported some plant and animal communities through a glaciation.

Paleoecology   The study of the relationships between prehistoric organisms and their physical environments.

Palynology   The study of fossil and modern pollen.

Parkland   A type of forest in which the trees are widely spaced over a landscape covered with herbaceous vegetation.

Periglacial environments   Environments at the immediate margins of glaciers and ice sheets, greatly influenced by the cold temperature of the ice.

Permafrost   Permanently frozen ground, found in arctic, subarctic, and alpine regions.

Photosynthesis   The synthesis by plants of carbohydrates from carbon dioxide and water by means of chlorophyll, using light as a source of energy.

Piedmont glacier   A mountain glacier that flows out from the mountain front onto the adjacent plains (the piedmont region).

Pleistocene epoch   An epoch of the Quaternary Period, spanning the interval from 1.7 million years ago to 10,000 years ago. The Pleistocene is characterized by a series of major glaciations.

Precipitate   To separate from a solution by a chemical or physical change.

Prograde   To build a fan of sediments from a stream out into a lake.

Projectile points   Sharp, pointed heads of stone or other material, attached to a shaft to make a projectile that is thrown or shot as a weapon. These include spearheads, arrowheads, and darts.

Proxy data   In Quaternary studies, data from, for example, fossil organisms, sediments, and ice cores used to reconstruct past environments; proxy data serve as a substitute for direct measurements of such phenomena as past temperatures, precipitation, and sea level.

Quaternary Period   The second period of the Cenozoic era, following the Tertiary and spanning the interval from about 1.7 million years ago to the present.

Sorted stone polygon   A form of patterned ground, the mesh of which is mainly polygonal in shape, with a sorted appearance resulting from a border of stones surrounding finer material.

Stratigraphic column   The arrangement of layers of sediments (strata) in geographic position and chronologic sequence.

Symbiosis   The living together of two or more different species of organisms in a way that is mutually beneficial.

Taphonomy   The process by which organisms become preserved in the fossil record; also the study of that process.

Terminal moraine   The end moraine that marks the farthest advance of a glacier or ice sheet.

Treeline   The altitudinal limit of tree species in mountain regions and the latitudinal limit of tree species at high latitudes.

Unconsolidated sediments   Sediments with particles not cemented together or turned to stone.

Water-lain sediments   Sediments that are deposited in water.

Wisconsin glaciation   The last major glaciation in North America, spanning the interval from about 110,000 to 10,000 yr B.P.

INDEX

ACL4066

DATE DUE